Lab Workbook
Modern Welding

by

William A. Bowditch
Career Education Consultant
Portage, Michigan
Member, American Welding Society
Member, Association for Career and Technical Education

Kevin E. Bowditch
Welding Engineer Specialist
Subaru of Indiana Automotive Inc.
Lafayette, Indiana
Member, American Welding Society
Member, Association for Career and Technical Education

Mark A. Bowditch
Member, American Welding Society
Member, Association for Career and Technical Education

Publisher
The Goodheart-Willcox Company, Inc.
Tinley Park, IL
www.g-w.com

Copyright © 2013
by
The Goodheart-Willcox Company, Inc.

Previous editions copyright 2004, 2000, 1997, 1994, 1989

All rights reserved. No part of this work may be reproduced, stored, or transmitted in any form or by any electronic or mechanical means, including information storage and retrieval systems, without the prior written permission of The Goodheart-Willcox Company, Inc.

Manufactured in the United States of America.

ISBN 978-1-60525-797-6

6 7 8 9 – 13 – 17

The Goodheart-Willcox Company, Inc. Brand Disclaimer: Brand names, company names, and illustrations for products and services included in this text are provided for educational purposes only and do not represent or imply endorsement or recommendation by the author or the publisher.

The Goodheart-Willcox Company, Inc. Safety Notice: The reader is expressly advised to carefully read, understand, and apply all safety precautions and warnings described in this book or that might also be indicated in undertaking the activities and exercises described herein to minimize risk of personal injury or injury to others. Common sense and good judgment should also be exercised and applied to help avoid all potential hazards. The reader should always refer to the appropriate manufacturer's technical information, directions, and recommendations; then proceed with care to follow specific equipment operating instructions. The reader should understand these notices and cautions are not exhaustive.

The publisher makes no warranty or representation whatsoever, either expressed or implied, including but not limited to equipment, procedures, and applications described or referred to herein, their quality, performance, merchantability, or fitness for a particular purpose. The publisher assumes no responsibility for any changes, errors, or omissions in this book. The publisher specifically disclaims any liability whatsoever, including any direct, indirect, incidental, consequential, special, or exemplary damages resulting, in whole or in part, from the reader's use or reliance upon the information, instructions, procedures, warnings, cautions, applications, or other matter contained in this book. The publisher assumes no responsibility for the activities of the reader.

The Goodheart-Willcox Company, Inc. Internet Disclaimer: The Internet resources and listings in this Goodheart-Willcox Publisher product are provided solely as a convenience to you. These resources and listings were reviewed at the time of publication to provide you with accurate, safe, and appropriate information. Goodheart-Willcox Publisher has no control over the referenced websites and, due to the dynamic nature of the Internet, is not responsible or liable for the content, products, or performance of links to other websites or resources. Goodheart-Willcox Publisher makes no representation, either expressed or implied, regarding the content of these websites, and such references do not constitute an endorsement or recommendation of the information or content presented. It is your responsibility to take all protective measures to guard against inappropriate content, viruses, or other destructive elements

Introduction

The **Lab Workbook for Modern Welding** is intended to be used with the **Modern Welding** text. This manual will help you to practice the welding techniques for the variety of welding processes presented in the text. Answering questions in the various lessons will help you to ensure that you have mastered the technical knowledge presented in the text.

Safety is a very important aspect of welding and is thoroughly discussed in Chapter 1. Safety notes and cautions are also printed in red throughout the textbook wherever they apply. Prior to beginning any new welding process, you should read about the process in the text and then satisfactorily pass the safety test(s) for that process.

The **Modern Welding** textbook and lab workbook are divided into nine parts. Each part covers one or more welding processes. Part 1, *Welding Fundamentals*, covers safety, print reading, reading welding symbols, and welding and cutting processes. Part 2 covers the shielded metal arc welding (SMAW) processes. Part 3 deals with the gas tungsten arc welding (GTAW), gas metal arc welding (GMAW), and flux cored arc welding (FCAW) processes. Part 4 explains the plasma arc cutting and gouging process. Part 5 covers oxyfuel gas welding and cutting, soldering, brazing, and braze welding. Part 6 covers resistance welding processes. In Part 7, special processes of welding and cutting are explained. Part 8, *Metal Technology*, covers the production, properties, and heat treatment of metals. Weld inspection and testing, weld and welder qualifications, the welding shop, and the important skill of getting a job in the welding field are discussed in Part 9, *Professional Welding*.

Although your instructor may assign chapters out of sequence, it is important that you study Part 1 first. Part 1 (Chapters 1–4) should be studied first because it covers print reading and reading AWS welding symbols. All the Jobs in this manual are drawn using AWS symbols. Therefore, you must know how to read these drawings and welding symbols.

Each lesson includes objectives, section and heading references, and specific instructions. Several types of questions are used, including matching, true or false, identification of parts, multiple choice, fill-in, and written essay answers. Some of the questions involve math calculations, while others require the use of tables and charts. The questions are designed to test your technical knowledge of welding techniques. Do your best to answer these questions completely and accurately.

Most of the lessons include one or more Jobs. Read the Job directions carefully. Be certain to follow all of the safety precautions. Inspection criteria and weld test procedures will be given for each Job. Inspect each weld prior to submitting it to the instructor for grading.

It is strongly advised that you keep a record of your completed lessons and jobs on the "Lesson and Job Completion Record" on pages 9–14 of this lab workbook. In addition to keeping track of the completion of lab assignments, you can include this record with your résumé to help demonstrate your experience and knowledge in welding to an employer.

William A. Bowditch
Kevin E. Bowditch
Mark A. Bowditch

Table of Contents

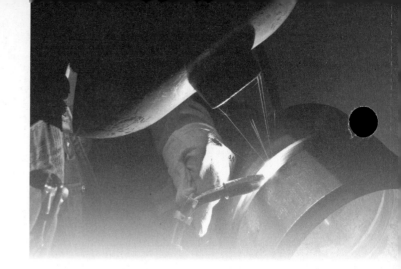

PART 1
Welding Fundamentals

Lesson 1A	General Shop Safety	15
Lesson 1B	Oxyfuel Gas Welding and Cutting Safety Test	17
Lesson 1C	Arc Welding and Cutting Safety	21
Lesson 1D	Resistance and Special Welding Processes Safety	23
Lesson 2	Print Reading	25
Lesson 3A	Welding Joints and Positions	29
Lesson 3B	Reading Welding Symbols	33
Lesson 3C	Electrode Angles	39
Lesson 4A	Welding and Cutting Processes	41
Lesson 4B	Welding and Cutting Processes	45
Lesson 4C	Welding and Cutting Processes	49

PART 2
Shielded Metal Arc Welding

Lesson 5A	SMAW Equipment and Supplies	51
Lesson 5B	SMAW Equipment and Supplies	55
Lesson 6A	Shielded Metal Arc Welding Safety	59
Lesson 6B	Shielded Metal Arc Welding Fundamentals	61
Job 6B-1	Inspecting and Setting Up an Arc Welding Station	65
Job 6B-2	Striking an Arc and Running Short Beads	66
Job 6B-3	Running Arc Beads	68
Job 6B-4	Running Weave Beads	69
Job 6B-5	Running Stringer Beads	70
Lesson 6C	SMAW in the Flat Position	71
Job 6C-1	Square-Groove Weld on an Edge Joint in the Flat Welding Position	75
Job 6C-2	Fillet Weld on a Lap Joint in the Flat Welding Position	76
Job 6C-3	Fillet Weld on a T-Joint in the Flat Welding Position	77
Job 6C-4	Bevel-Groove Weld on a Butt Joint in the Flat Welding Position	78
Lesson 6D	SMAW in the Horizontal Position	79
Job 6D-1	Fillet Weld on a Lap Joint in the Horizontal Welding Position	81
Job 6D-2	Fillet Weld on a T-Joint in the Horizontal Welding Position	82
Job 6D-3	Bevel-Groove Weld on a Butt Joint and a V-Groove Weld on an Outside Corner Joint in the Horizontal Welding Position	83

Lesson 6E	SMAW in the Vertical and Overhead Positions	85
Job 6E-1	Fillet Weld on a Lap Joint in the Vertical Welding Position	87
Job 6E-2	Fillet Weld on a T-Joint in the Vertical Welding Position	88
Job 6E-3	V-Groove Weld on a Butt Joint in the Vertical Welding Position	89
Job 6E-4	Fillet Weld on a Lap Joint in the Overhead Welding Position	90
Job 6E-5	Fillet Weld on a T-Joint in the Overhead Welding Position	91
Job 6E-6	Bevel-Groove Weld on a Butt Joint in the Overhead Welding Position	92

PART 3
Gas Tungsten, Gas Metal, and Flux Cored Arc Welding

Lesson 7A	GTAW Equipment and Supplies	93
Lesson 7B	GMAW and FCAW Equipment and Supplies	99
Lesson 8A	Gas Tungsten Arc Welding Safety	103
Lesson 8B	Gas Tungsten Arc Welding Principles	105
Lesson 8C	Gas Tungsten Arc Welding Procedures	109
Job 8C-1	Fillet Weld on a Lap Joint in the Flat Welding Position	113
Job 8C-2	Fillet Weld on an Inside Corner in the Flat Welding Position	115
Job 8C-3	Square-Groove Weld on a Butt Joint in the Flat Welding Position	117
Job 8C-4	Fillet Weld on a Lap Joint in the Horizontal Welding Position	119
Job 8C-5	Fillet Weld on an Inside Corner and a T-Joint in the Horizontal Welding Position	121
Job 8C-6	Square-Groove Weld on a Butt Joint in the Horizontal Welding Position	123
Job 8C-7	Fillet Weld on a Lap Joint in the Vertical Welding Position	125
Job 8C-8	Fillet Weld on a T-Joint in the Vertical Welding Position	127
Job 8C-9	Square-Groove Weld on a Butt Joint in the Vertical Welding Position	129
Job 8C-10	Fillet Weld on a Lap Joint in the Overhead Welding Position	131
Job 8C-11	Fillet Weld on a T-Joint in the Overhead Welding Position	133
Job 8C-12	Square-Groove Weld on a Butt Joint in the Overhead Welding Position	135
Job 8C-13	Groove Welds on Stainless Steel	137
Lesson 9A	Gas Metal and Flux Cored Arc Welding Safety	139
Lesson 9B	Gas Metal and Flux Cored Arc Welding Principles	141
Lesson 9C	Gas Metal and Flux Cored Arc Welding— Setting Up the Station	145
Lesson 9D	GMAW and FCAW Procedures for Flat and Horizontal Welds	151
Job 9D-1	Adjusting GMAW and FCAW Welding Machines	155
Job 9D-2	Setting a GMAW Welding Machine and Making a Fillet Weld on a Lap Joint in the Flat Welding Position	157
Job 9D-3	Fillet Weld on an Inside Corner and a T-Joint in the Flat Welding Position	159
Job 9D-4	Square-Groove Weld on a Butt Joint in the Flat Welding Position	161
Job 9D-5	Fillet Weld on a Lap Joint in the Horizontal Welding Position	163
Job 9D-6	Fillet Weld on a T-Joint in the Horizontal Welding Position	165
Job 9D-7	Square-Groove Weld on a Butt Joint in the Horizontal Welding Position	167
Lesson 9E	GMAW and FCAW Procedures for Vertical and Overhead Welds	169
Job 9E-1	Fillet Weld on a Lap Joint in the Vertical Welding Position	171
Job 9E-2	Fillet Weld on an Inside Corner and a T-Joint in the Vertical Welding Position	173
Job 9E-3	Bevel-Groove Weld on a Butt Joint in the Vertical Welding Position	175
Job 9E-4	Fillet Weld on a Lap Joint in the Overhead Welding Position	177
Job 9E-5	Fillet Weld on a T-Joint in the Overhead Welding Position	179
Job 9E-6	Square-Groove Weld on a Butt Joint in the Overhead Welding Position	181

PART 4
Plasma Arc Cutting

Lesson 10   Plasma Arc Cutting...183
    Job 10-1   Piercing and Cutting Using the Plasma Arc Cutting Process 185
    Job 10-2   Cutting a Shape Using the Plasma Arc Cutting Process 187

PART 5
Oxyfuel Gas Processes

Lesson 11A   Oxyfuel Gas Welding Equipment and Supplies 189
Lesson 11B   Oxyfuel Gas Welding and Cutting Safety..193
Lesson 12A   Oxyfuel Gas Welding—Turning the Outfit On and Off...............................197
    Job 12A-1   Turning On, Lighting, and Shutting Down an Oxyacetylene Welding Outfit............201
Lesson 12B   Oxyfuel Gas Welding—Running a Continuous Weld Pool 203
    Job 12B-1   Running a Continuous Weld Pool...205
    Job 12B-2   Square-Groove Weld on an Outside Corner.. 206
Lesson 12C   Oxyfuel Gas Welding—Welding Mild Steel in the Flat Welding Position 207
    Job 12C-1   Fillet Weld on a Lap Joint in the Flat Welding Position 209
    Job 12C-2   Fillet Weld on an Inside Corner in the Flat Welding Position 210
    Job 12C-3   Square-Groove Weld on an Outside Corner in the Flat Welding Position 211
    Job 12C-4   Square-Groove Weld on a Butt Joint in the Flat Welding Position....................212
    Job 12C-5   Fillet Weld on an Inside Corner in the Flat Welding Position 213
    Job 12C-6   Fillet Weld on a Lap Joint in the Flat Welding Position 214
    Job 12C-7   V-Groove Weld on a Butt Joint in the Flat Welding Position 215
Lesson 12D   Oxyfuel Gas Welding—Welding Mild Steel in the Horizontal Welding Position 217
    Job 12D-1   Fillet Weld on a Lap Joint in the Horizontal Welding Position 219
    Job 12D-2   Fillet Weld on a T-Joint in the Horizontal Welding Position.......................220
    Job 12D-3   Square-Groove Weld on a Butt Joint in the Horizontal Welding Position..........221
Lesson 12E   Oxyfuel Gas Welding—Welding Mild Steel in the Vertical Welding Position..........223
    Job 12E-1   Fillet Weld on a Lap Joint in the Vertical Welding Position.......................227
    Job 12E-2   Fillet Weld on a T-Joint in the Vertical Welding Position 228
    Job 12E-3   V-Groove Weld on an Outside Corner in the Vertical Welding Position 229
Lesson 12F   Oxyfuel Gas Welding—Welding Mild Steel in the Overhead Welding Position..........231
    Job 12F-1   Square-Groove Weld on a Butt Joint, Fillet Weld on a Lap Joint, and
                   Fillet Weld on a T-Joint in the Overhead Welding Position.......................233
Lesson 13   Oxyfuel Gas Cutting Equipment and Supplies 235
Lesson 14   Oxyfuel Gas Cutting—Cutting Steel.. 239
    Job 14-1   Cutting Mild Steel with an Oxyfuel Gas Torch 243
    Job 14-2   Cutting Shapes with an Oxyfuel Gas Torch...244
    Job 14-3   Removing Weld Reinforcement from the Face of a Weld.........................245
    Job 14-4   Setting Up and Inspecting a Semiautomatic Oxyfuel Gas Cutting Torch 246
    Job 14-5   Making Straight-Line Cuts on Steel to Produce Square and Beveled Edges..........247
Lesson 15   Soldering.. 249
    Job 15-1   Soldering Copper Fittings Used in Plumbing 253
    Job 15-2   Soldering a Folded Seam..254
Lesson 16A   Brazing and Braze Welding Principles 255
    Job 16A-1   Braze Welding a Butt Joint, a Lap Joint, and a T-Joint in the Flat
                   Welding Position..257
Lesson 16B   Brazing and Braze Welding Processes..259
    Job 16B-1   Braze Welding a V-Groove on an Outside Corner in the Flat and
                   Horizontal Welding Positions...261

PART 6
Resistance Welding

Lesson 17A	Resistance Welding Safety	263
Lesson 17B	Resistance Welding Machines	265
Job 17B-1	Major Components of Resistance Spot Welding Machines	269
Lesson 17C	Resistance Welding Electrical Components	271
Job 17C-1	Resistance Spot Welding Machine Cooling and Electrodes	273
Lesson 18	Resistance Welding	277
Job 18-1	Setting the Variables and Making Spot Welds	281
Job 18-2	Setting the Variables and Making Spot Welds	285

PART 7
Special Processes

Lesson 19A	Arc-Related Welding Processes	289
Lesson 19B	Solid-State and Other Welding Processes	293
Lesson 20	Special Ferrous Welding Applications	297
Job 20-1	Square-Groove Weld on a Butt Joint on Stainless Steel Using SMAW in the Flat Welding Position	301
Job 20-2	Fillet Weld on a T-Joint on Stainless Steel Using SMAW in the Flat Welding Position	303
Job 20-3	Square-Groove Weld on a Butt Joint and a Fillet Weld in the Flat Welding Position	304
Job 20-4	Square or V-Groove Weld on a Butt Joint on Cast Iron Using SMAW	306
Lesson 21	Nonferrous Welding Applications	307
Job 21-1	Groove Welds and Fillet Welds on Aluminum Using GTAW	311
Lesson 22	Pipe and Tube Welding	313
Job 22-1	Welding Mild Steel Pipe in the 1G Position Using SMAW	317
Job 22-2	Welding Mild Steel Pipe in the 2G and 5G Positions Using SMAW	319
Job 22-3	Welding Mild Steel Pipe in the 5G Position Using GTAW	321
Job 22-4	Welding Mild Steel Pipe in the 2G and 5G Positions Using GMAW	323
Job 22-5	Welding Mild Steel Pipe in the 6G Position Using GMAW	325
Job 22-6	Welding Mild Steel Pipe in the 6G Position Using SMAW	327
Lesson 23A	Special Cutting Safety	329
Lesson 23B	Special Cutting Processes	331
Job 23B-1	Inspecting a CAC-A Welding Station	335
Job 23B-2	Cutting and Piercing Using CAC-A	336
Job 23B-3	Removing a Weld or Weld Reinforcement Using CAC-A	337
Lesson 24	Underwater Welding and Cutting	339
Lesson 25	Automatic and Robotic Welding	343
Lesson 26	Metal Surfacing	347
Job 26-1	Hardfacing a Steel Plate Using the Oxyfuel Gas Process	351

PART 8
Metal Technology

Lesson 27	Production of Metals	353
Lesson 28	Metal Properties and Identification	357
Job 28-1	Identification of Metals	361
Lesson 29	Heat Treatment of Metals	363
Job 29-1	Heat Treating a Cold Chisel	367

PART 9
Professional Welding

Lesson 30	Inspecting and Testing Welds	369
Job 30-1	Magnetic Particle Inspection	373
Job 30-2	Liquid Penetrant Inspection	375
Job 30-3	Guided Bend Tests	377
Job 30-4	Tensile Testing	379
Job 30-5	Hardness Testing	381
Lesson 31	Procedure and Welder Qualifications	383
Job 31-1	Writing a Welding Procedure Specification and Qualifying the WPS	385
Job 31-2	Fillet Weld Performance Test	390
Lesson 32	The Welding Shop	391
Lesson 33	Getting and Holding a Job in the Welding Industry	393
Lesson 34	Technical Data	397

Lesson and Job Completion Record

Name _____   Course _____   Class _____
School _____   Instructor _____

As stated in the Introduction, the following chart is provided to assist you in keeping accurate records of all lab assignments you successfully complete.

Completion

PART 1
Welding Fundamentals

Lesson 1A	General Shop Safety	_____
Lesson 1B	Oxyfuel Gas Welding and Cutting Safety	_____
Lesson 1C	Arc Welding and Cutting Safety	_____
Lesson 1D	Resistance and Special Welding Processes Safety	_____
Lesson 2	Print Reading	_____
Lesson 3A	Welding Joints and Positions	_____
Lesson 3B	Reading Welding Symbols	_____
Lesson 3C	Electrode Angles	_____
Lesson 4A	Welding and Cutting Processes	_____
Lesson 4B	Welding and Cutting Processes	_____
Lesson 4C	Welding and Cutting Processes	_____

PART 2
Shielded Metal Arc Welding

Lesson 5A	SMAW Equipment and Supplies	_____
Lesson 5B	SMAW Equipment and Supplies	_____
Lesson 6A	Shielded Metal Arc Welding Safety	_____
Lesson 6B	Shielded Metal Arc Welding Fundamentals	_____
Job 6B-1	Inspecting and Setting Up an Arc Welding Station	_____
Job 6B-2	Striking an Arc and Running Short Beads	_____
Job 6B-3	Running Arc Beads	_____
Job 6B-4	Running Weave Beads	_____
Job 6B-5	Running Stringer Beads	_____
Lesson 6C	SMAW in the Flat Position	_____
Job 6C-1	Square-Groove Weld on an Edge Joint in the Flat Welding Position	_____
Job 6C-2	Fillet Weld on a Lap Joint in the Flat Welding Position	_____
Job 6C-3	Fillet Weld on a T-Joint in the Flat Welding Position	_____
Job 6C-4	Bevel-Groove Weld on a Butt Joint in the Flat Welding Position	_____

Completion

Lesson 6D	SMAW in the Horizontal Position	_____
Job 6D-1	Fillet Weld on a Lap Joint in the Horizontal Welding Position	_____
Job 6D-2	Fillet Weld on a T-Joint in the Horizontal Welding Position	_____
Job 6D-3	Bevel-Groove Weld on a Butt Joint and a V-Groove Weld on an Outside Corner Joint in the Horizontal Welding Position	_____
Lesson 6E	SMAW in the Vertical and Overhead Positions	_____
Job 6E-1	Fillet Weld on a Lap Joint in the Vertical Welding Position	_____
Job 6E-2	Fillet Weld on a T-Joint in the Vertical Welding Position	_____
Job 6E-3	V-Groove Weld on a Butt Joint in the Vertical Welding Position	_____
Job 6E-4	Fillet Weld on a Lap Joint in the Overhead Welding Position	_____
Job 6E-5	Fillet Weld on a T-Joint in the Overhead Welding Position	_____
Job 6E-6	Bevel-Groove Weld on a Butt Joint in the Overhead Welding Position	_____

PART 3
Gas Tungsten, Gas Metal, and Flux Cored Arc Welding

Lesson 7A	GTAW Equipment and Supplies	_____
Lesson 7B	GMAW and FCAW Equipment and Supplies	_____
Lesson 8A	Gas Tungsten Arc Welding Safety	_____
Lesson 8B	Gas Tungsten Arc Welding Principles	_____
Lesson 8C	Gas Tungsten Arc Welding Procedures	_____
Job 8C-1	Fillet Weld on a Lap Joint in the Flat Welding Position	_____
Job 8C-2	Fillet Weld on an Inside Corner in the Flat Welding Position	_____
Job 8C-3	Square-Groove Weld on a Butt Joint in the Flat Welding Position	_____
Job 8C-4	Fillet Weld on a Lap Joint in the Horizontal Welding Position	_____
Job 8C-5	Fillet Weld on an Inside Corner and a T-Joint in the Horizontal Welding Position	_____
Job 8C-6	Square-Groove Weld on a Butt Joint in the Horizontal Welding Position	_____
Job 8C-7	Fillet Weld on a Lap Joint in the Vertical Welding Position	_____
Job 8C-8	Fillet Weld on a T-Joint in the Vertical Welding Position	_____
Job 8C-9	Square-Groove Weld on a Butt Joint in the Vertical Welding Position	_____
Job 8C-10	Fillet Weld on a Lap Joint in the Overhead Welding Position	_____
Job 8C-11	Fillet Weld on a T-Joint in the Overhead Welding Position	_____
Job 8C-12	Square-Groove Weld on a Butt Joint in the Overhead Welding Position	_____
Job 8C-13	Groove Welds on Stainless Steel	_____
Lesson 9A	Gas Metal and Flux Cored Arc Welding Safety	_____
Lesson 9B	Gas Metal and Flux Cored Arc Welding Principles	_____

Name _____

Completion

Lesson 9C	Gas Metal and Flux Cored Arc Welding—Setting Up the Station	_____
Lesson 9D	GMAW and FCAW Procedures for Flat and Horizontal Welds	_____
Job 9D-1	Adjusting GMAW and FCAW Welding Machines	_____
Job 9D-2	Setting a GMAW Welding Machine and Making a Fillet Weld on a Lap Joint in the Flat Welding Position	_____
Job 9D-3	Fillet Weld on an Inside Corner and a T-Joint in the Flat Welding Position	_____
Job 9D-4	Square-Groove Weld on a Butt Joint in the Flat Welding Position	_____
Job 9D-5	Fillet Weld on a Lap Joint in the Horizontal Welding Position	_____
Job 9D-6	Fillet Weld on a T-Joint in the Horizontal Welding Position	_____
Job 9D-7	Square-Groove Weld on a Butt Joint in the Horizontal Welding Position	_____
Lesson 9E	GMAW and FCAW Procedures for Vertical and Overhead Welds	_____
Job 9E-1	Fillet Weld on a Lap Joint in the Vertical Welding Position	_____
Job 9E-2	Fillet Weld on an Inside Corner and a T-Joint in the Vertical Welding Position	_____
Job 9E-3	Bevel-Groove Weld on a Butt Joint in the Vertical Welding Position	_____
Job 9E-4	Fillet Weld on a Lap Joint in the Overhead Welding Position	_____
Job 9E-5	Fillet Weld on a T-Joint in the Overhead Welding Position	_____
Job 9E-6	Square-Groove Weld on a Butt Joint in the Overhead Welding Position	_____

PART 4
Plasma Arc Cutting

Lesson 10	Plasma Arc Cutting	_____
Job 10-1	Piercing and Cutting Using the Plasma Arc Cutting Process	_____
Job 10-2	Cutting a Shape Using the Plasma Arc Cutting Process	_____

PART 5
Oxyfuel Gas Processes

Lesson 11A	Oxyfuel Gas Welding Equipment and Supplies	_____
Lesson 11B	Oxyfuel Gas Welding and Cutting Safety	_____
Lesson 12A	Oxyfuel Gas Welding—Turning the Outfit On and Off	_____
Job 12A-1	Turning On, Lighting, and Shutting Down an Oxyacetylene Welding Outfit	_____
Lesson 12B	Oxyfuel Gas Welding—Running a Continuous Weld Pool	_____
Job 12B-1	Running a Continuous Weld Pool	_____
Job 12B-2	Square-Groove Weld on an Outside Corner	_____

		Completion
Lesson 12C	Oxyfuel Gas Welding—Welding Mild Steel in the Flat Welding Position	_____
Job 12C-1	Fillet Weld on a Lap Joint in the Flat Welding Position	_____
Job 12C-2	Fillet Weld on an Inside Corner in the Flat Welding Position	_____
Job 12C-3	Square-Groove Weld on an Outside Corner in the Flat Welding Position	_____
Job 12C-4	Square-Groove Weld on a Butt Joint in the Flat Welding Position	_____
Job 12C-5	Fillet Weld on an Inside Corner in the Flat Welding Position	_____
Job 12C-6	Fillet Weld on a Lap Joint in the Flat Welding Position	_____
Job 12C-7	V-Groove Weld on a Butt Joint in the Flat Welding Position	_____
Lesson 12D	Oxyfuel Gas Welding—Welding Mild Steel in the Horizontal Welding Position	_____
Job 12D-1	Fillet Weld on a Lap Joint in the Vertical Welding Position	_____
Job 12D-2	Fillet Weld on a T-Joint in the Horizontal Welding Position	_____
Job 12D-3	Square-Groove Weld on a Butt Joint in the Horizontal Welding Position	_____
Lesson 12E	Oxyfuel Gas Welding—Welding Mild Steel in the Vertical Welding Position	_____
Job 12E-1	Fillet Weld on a Lap Joint in the Vertical Welding Position	_____
Job 12E-2	Fillet Weld on a T-Joint in the Vertical Welding Position	_____
Job 12E-3	V-Groove Weld on an Outside Corner in the Vertical Welding Position	_____
Lesson 12F	Oxyfuel Gas Welding—Welding Mild Steel in the Overhead Welding Position	_____
Job 12F-1	Square-Groove Weld on a Butt Joint, Fillet Weld on a Lap Joint, and Fillet Weld on a T-Joint in the Overhead Welding Position	_____
Lesson 13	Oxyfuel Gas Cutting Equipment and Supplies	_____
Lesson 14	Oxyfuel Gas Cutting—Cutting Steel	_____
Job 14-1	Cutting Mild Steel with an Oxyfuel Gas Torch	_____
Job 14-2	Cutting Shapes with an Oxyfuel Gas Torch	_____
Job 14-3	Removing Weld Reinforcement from the Face of a Weld	_____
Job 14-4	Setting Up and Inspecting a Semiautomatic Oxyfuel Gas Cutting Torch	_____
Job 14-5	Making Straight-Line Cuts on Steel to Produce Square and Beveled Edges	_____
Lesson 15	Soldering	_____
Job 15-1	Soldering Copper Fittings Used in Plumbing	_____
Job 15-2	Soldering a Folded Seam	_____
Lesson 16A	Brazing and Braze Welding Principles	_____
Job 16A-1	Braze Welding a Butt Joint, a Lap Joint, and a T-Joint in the Flat Welding Position	_____
Lesson 16B	Brazing and Braze Welding Processes	_____
Job 16B-1	Braze Welding a V-Groove on an Outside Corner in the Flat and Horizontal Welding Positions	_____

Name _____

Completion

PART 6
Resistance Welding

Lesson 17A	Resistance Welding Safety	_____
Lesson 17B	Resistance Welding Machines	_____
Job 17B-1	Major Components of Resistance Spot Welding Machines	_____
Lesson 17C	Resistance Welding Electrical Components	_____
Job 17C-1	Resistance Spot Welding Machine Cooling and Electrodes	_____
Lesson 18	Resistance Welding	_____
Job 18-1	Setting the Variables and Making Spot Welds	_____
Job 18-2	Setting the Variables and Making Spot Welds	_____

PART 7
Special Processes

Lesson 19A	Arc-Related Welding Processes	_____
Lesson 19B	Solid-State and Other Welding Processes	_____
Lesson 20	Special Ferrous Welding Applications	_____
Job 20-1	Square-Groove Weld on a Butt Joint on Stainless Steel Using SMAW in the Flat Welding Position	_____
Job 20-2	Fillet Weld on a T-Joint on Stainless Steel Using SMAW in the Flat Welding Position	_____
Job 20-3	Square-Groove Weld on a Butt Joint and a Fillet Weld in the Flat Welding Position	_____
Job 20-4	Square or V-Groove Weld on a Butt Joint on Cast Iron Using SMAW	_____
Lesson 21	Nonferrous Welding Applications	_____
Job 21-1	Groove Welds and Fillet Welds on Aluminum Using GTAW	_____
Lesson 22	Pipe and Tube Welding	_____
Job 22-1	Welding Mild Steel Pipe in the 1G Position Using SMAW	_____
Job 22-2	Welding Mild Steel Pipe in the 2G and 5G Positions Using SMAW	_____
Job 22-3	Welding Mild Steel Pipe in the 5G Position Using GTAW	_____
Job 22-4	Welding Mild Steel Pipe in the 2G and 5G Positions Using GMAW	_____
Job 22-5	Welding Mild Steel Pipe in the 6G Position Using GMAW	_____
Job 22-6	Welding Mild Steel Pipe in the 6G Position Using SMAW	_____
Lesson 23A	Special Cutting Safety	_____
Lesson 23B	Special Cutting Processes	_____
Job 23B-1	Inspecting a CAC-A Welding Station	_____
Job 23B-2	Cutting and Piercing Using CAC-A	_____
Job 23B-3	Removing a Weld or Weld Reinforcement Using CAC-A	_____

                                                                                            Completion

    Lesson 24   Underwater Welding and Cutting   _____

    Lesson 25   Automatic and Robotic Welding   _____

    Lesson 26   Metal Surfacing   _____

        Job 26-1   Hardfacing a Steel Plate Using the Oxyfuel Gas Process   _____

PART 8
Metal Technology

    Lesson 27   Production of Metals   _____

    Lesson 28   Metal Properties and Identification   _____

        Job 28-1   Identification of Metals   _____

    Lesson 29   Heat Treatment of Metals   _____

        Job 29-1   Heat Treating a Cold Chisel   _____

PART 9
Professional Welding

    Lesson 30   Inspecting and Testing Welds   _____

        Job 30-1   Magnetic Particle Inspection   _____

        Job 30-2   Liquid Penetrant Inspection   _____

        Job 30-3   Guided Bend Tests   _____

        Job 30-4   Tensile Testing   _____

        Job 30-5   Hardness Testing   _____

    Lesson 31   Procedure and Welder Qualifications   _____

        Job 31-1   Writing a Welding Procedure Specification and Qualifying the WPS   _____

        Job 31-2   Fillet Weld Performance Test   _____

    Lesson 32   The Welding Shop   _____

    Lesson 33   Getting and Holding a Job in the Welding Industry   _____

    Lesson 34   Technical Data   _____

Lesson 1A

General Shop Safety

Name _____ Date _____
Class _____ Instructor _____

Learning Objective
You will be able to identify several of the safety hazards and precautions required when working in any typical shop area.

Instructions
Carefully read Headings 1.1 through 1.8.3 and study the figures in Chapter 1 of the text. Then answer the following questions.

1. *True or False?* Your age, state of health, job skills, and attitude have little to do with how you feel about yourself or your job.

    1. _____

2. *True or False?* Health is *not* considered a physical factor in accidents.

    2. _____

3. *True or False?* Paint, oil, and cleaning chemicals should be stored in a steel cabinet for safety.

    3. _____

4. *True or False?* A person who is under stress may be distracted by worry, anger, sorrow, or thoughts of love.

    4. _____

5. *True or False?* Fire extinguishers and fire blankets are mounted on a surface that is painted a bright yellow.

    5. _____

6. *True or False?* It is good safety practice to clearly mark voltages of 220V or higher with danger signs.

    6. _____

7. *True or False?* When lifting, it is a good practice to wear a back brace, lift with your legs, and keep your back straight.

    7. _____

8. *True or False?* All hand tools should be inspected for worn power cords and loose parts.

    8. _____

9. *True or False?* Safety hazards, locations of fire extinguishers, proper sequence of operation, the proper placement of hands and feet, and machine safety features should be covered during job or task training.

    9. _____

Modern Welding Lab Workbook

10. *True or False?* Thinking about pleasant things to pass the time will help to promote safety in the shop.

    10. _____

11. *True or False?* An air-purifying respirator will filter dust and other particles from the air, but it does not protect against toxic fumes such as cadmium.

    11. _____

12. *True or False?* Fire extinguisher locations are painted yellow and black.

    12. _____

13. _____ is considered an important factor in shop safety.
    A. Housekeeping
    B. A good attitude
    C. The proper storage of hazardous materials
    D. Adequate job training
    E. All of the above.

    13. _____

14. Which of the following line voltages need *not* be marked "danger" or "danger high voltage"?
    A. 115 volts
    B. 230 volts
    C. 220 volts
    D. 440 volts
    E. All of the above.

    14. _____

15. *True or False?* You should bend your back slightly when lifting.

    15. _____

16. Well-designed welding shops have _____ ceilings to improve ventilation.

    16. _____

17. Many companies require a worker to wear a(n) _____ when lifting anything.

    17. _____

18. Dangerous corners or stairway overhangs should be painted with alternating yellow and _____ angled stripes for high visibility.

    18. _____

19. Heat _____ is characterized by faintness, dizziness, and heavy sweating.

    19. _____

20. Heat _____ can lead to unconsciousness or death.

    20. _____

Lesson 1B

Oxyfuel Gas Welding and Cutting Safety Test

Name _____ Date _____

Class _____ Instructor _____

> **Learning Objective**
> - You will be able to list several of the hazards and safety precautions to observe when working with oxyfuel gas welding and cutting equipment.
>
> **Instructions**
> *Carefully read Headings 1.2 through 1.4 and their subheadings. Also study Figures 1-7 through 1-21 in the text. Then answer the following questions.*

1. The filter lens recommended to be worn for most oxyfuel gas welding or cutting is a number ____.
   A. 1–2
   B. 3–6
   C. 7–9
   D. 10–12
   E. 13–15

   1. _____

2. The recommended fabric for clothing worn while oxyfuel gas welding or cutting is ____.
   A. wool
   B. flame-retardant-treated cotton
   C. leather
   D. Any of the above.
   E. None of the above.

   2. _____

3. Metal that has been left somewhere to be cooled should be clearly marked "____."

   3. _____

4. Acetylene gas should never be used at a pressure above ____.
   A. 10 psig
   B. 14 psig
   C. 14.7 psig
   D. 15 psig

   4. _____

Copyright by The Goodheart-Willcox Co., Inc.   Modern Welding Lab Workbook   17

5. Which of the following pieces of safety equipment must be used when welding in the presence of toxic fumes, smoke, or a lack of oxygen?
   A. A welding helmet equipped with a device that detects poisonous gas in the air.
   B. An air-purifying respirator.
   C. An air-supplied respirator.
   D. An OSHA-approved ventilation system capable of filtering dust, dirt, and metal particles.

5. _____

6. If the highest pressure reading on the face of a gauge is 200 psig, what is the highest working pressure at which this gauge should be used?
   A. 50 psig
   B. 100 psig
   C. 150 psig
   D. 175 psig
   E. 200 psig

6. _____

7. _____ may form toxic or deadly fumes when heated or welding.
   A. Beryllium
   B. Chromium
   C. Cadmium
   D. Lead
   E. All of the above.

7. _____

8. Oxygen and acetylene cylinders should always be _____.
   A. fastened to a wall or column
   B. moved with the safety cap in place
   C. stored upright in a special room or cage
   D. secured with sturdy chains or steel bands
   E. All of the above.

8. _____

9. *True or False?* Shirts with open pockets and trousers without cuffs should be worn when welding.

9. _____

10. *True or False?* Leather gauntlet-type gloves are best for oxyfuel gas welding in the overhead position.

10. _____

11. *True or False?* Ultraviolet and infrared rays are created when oxyfuel gas welding.

11. _____

12. *True or False?* An oxyfuel gas flame should be lighted with a safety match.

12. _____

13. *True or False?* The American Welding Society recommends that a label be attached to the major equipment in an oxyfuel gas welding station to warn of potential hazards.

13. _____

14. *True or False?* All gases should be bled from the torch and hoses when the station is shut down.

14. _____

Lesson 1B   Oxyfuel Gas Welding and Cutting Safety Test

Name _____

15. *True or False?* Whenever the welding torch is not in the welder's hand, it must be turned off.   15. _____

16. *True or False?* An oxyfuel gas cutting area must be cleared of combustible materials and at least one fire watch must be posted with a fire extinguisher.   16. _____

17. *True or False?* Sodium cyanide compounds can create toxic or deadly fumes when welded.   17. _____

18. To avoid possible burns, _____ or _____ should be used to pick up hot metal in the welding shop.   18. _____

19. The pressure regulator on a welding station is OFF when the regulator's adjusting screw is turned all the way _____. (in/out)   19. _____

20. A Type "C" fire extinguisher is used to extinguish _____.
    A. combustible liquid fires
    B. electrical fires
    C. solid material fires
    D. combustible metal fires   20. _____

Lesson 1C

Arc Welding and Cutting Safety

Name _____ Date _____

Class _____ Instructor _____

Learning Objective
- You will be able to list several hazards and safety precautions to observe when working with arc welding and cutting equipment.

Instructions
Carefully read Heading 1.5 and its subheadings and study Figures 1-22 and 1-23 in the text. Then answer the following questions.

1. *True or False?* Ultraviolet and infrared rays are created during arc welding or cutting.

   1. _____

2. *True or False?* High-pressure gas cylinders can be moved using a cylinder truck or by slightly tipping the cylinder and rolling it on its edge.

   2. _____

3. *True or False?* The American Welding Society recommends that a label be attached to all major equipment in an arc welding or cutting station to warn of dangers.

   3. _____

4. *True or False?* It is recommended that an arc welding station inspection be made with the power on.

   4. _____

5. *True or False?* Toxic fumes given off during arc welding should always be removed at a point above the welder's head.

   5. _____

6. *True or False?* Cadmium, chromium, beryllium, zinc, and lead can give off toxic or deadly fumes when welded or cut.

   6. _____

7. *True or False?* It is recommended that a filter-type respirator be worn when welding or cutting on materials that may produce toxic or poisonous fumes.

   7. _____

8. When arc cutting is being performed, it is recommended that a fire watch who has ready access to a(n) _____ be posted.

   8. _____

Copyright by The Goodheart-Willcox Co., Inc.   Modern Welding Lab Workbook   21

9. When it is necessary to run the welding leads across an aisle or high-traffic area, you should ____.
   A. lay the leads under a piece of channel iron
   B. lay the leads under a piece of angle iron
   C. Either A or B.
   D. Neither A nor B.

9. _____

10. First aid for a first degree burn involves applying ____ to the burn area.
    A. ice
    B. cold water
    C. cold compresses
    D. Either A or B.
    E. Either B or C.

10. _____

Lesson 1D

Resistance and Special Welding Processes Safety

Name _____ Date _____
Class _____ Instructor _____

Learning Objective
- You will be able to identify several of the hazards and safety precautions to observe when working with resistance and some other welding processes.

Instructions
Carefully read Headings 1.6 through 1.8.3 in the text. Then answer the following questions.

1. *True or False?* An unbreakable plastic face shield can be worn during resistance spot or seam welding.

    1. _____

2. *True or False?* It is *not* required for the welder to wear gloves while resistance welding.

    2. _____

3. *True or False?* You should *not* carry matches, butane lighters, or plastic combs in your pockets when resistance welding.

    3. _____

4. *True or False?* Hot sparks and molten metal may fly about during resistance spot welding.

    4. _____

5. *True or False?* While resistance spot welding, it is *not* a good safety practice to stand on an insulated platform or flooring material.

    5. _____

6. Special filter lenses that filter more than ultraviolet and infrared rays are required for ____ welding and ____ welding.

    6. _____

7. A welding operator should not open the control panels of resistance welding machines because of the ____ within the control cabinets.

    7. _____

8. Why is it dangerous to stand within a robot's working volume?
   _____
   _____
   _____

9. *True or False?* Explosive welding requires special training. 9. _____

10. *True or False?* Explosive welding and plasma arc welding are two welding processes that produce sounds so loud that they could damage the welder's hearing. 10. _____

Lesson 2

Print Reading

Name _____ Date _____
Class _____ Instructor _____

Learning Objective
You will be able to identify the various views in a sketch or drawing. You will also be able to follow a line or point from one view to another and find the sizes of various parts from a print.

Instructions
Carefully read Headings 2.1 through 2.6 of the text. Also study Figures 2-1 through 2-14. Then answer the following questions.

Use the following drawing when answering Questions 1–10.

Identify the ends of the lines shown on the drawing from lines already identified in other views. In question 1, identify the missing point 1. In question 2, identify point 2. Remember, when a line (s-t) appears as a point, it is identified as (s,t). The first letter written is the point closest to you: "s" is closest and "t" is farthest away.

1. _____
2. _____
3. _____
4. _____
5. _____
6. _____
7. _____
8. _____
9. _____

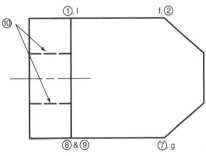

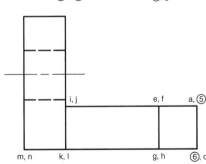

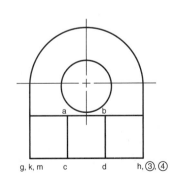

10. What do these hidden lines represent? _____

Modern Welding Lab Workbook   25

Use the following drawing when answering Questions 11–20.

What is the distance between arrowheads or the size of the following numbers on the drawing?

11. _____
12. _____
13. _____
14. _____
15. _____
16. _____
17. _____
18. _____
19. _____
20. _____

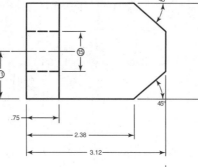
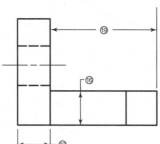
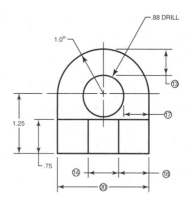

Use the views lettered A, B, and C in the preceding drawing to answer Questions 21 through 25.

21. What view is view "A"?   21. _____

22. What view is view "B"?   22. _____

23. What view is view "C"?   23. _____

24. What method of projection was used on the drawings?   24. _____

25. Are these drawings sketches or mechanical drawings?   25. _____

26. What is the name of the mathematical formula used to determine the length of one side of a triangle when the length of the other two sides is known?   26. _____

27. Which type of axonometric view shows all three angles and corners of equal size?   27. _____
    A. isometric
    B. diametric
    C. trimetric

28. Which of the following can be determined by using the formula 1/2 × d?   28. _____
    A. diameter
    B. radius
    C. perimeter
    D. circumference

Name _____

29. Unlike axonometric and orthographic (or multiview) projection, _____ projection uses parallel projection rays that are not perpendicular to the viewing plane.

29. _____

30. A(n) _____ view depicts a cross section of an object in an imaginary cut plane.

30. _____

Lesson 3A

Welding Joints and Positions

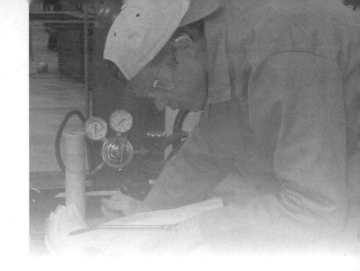

Name _____ Date _____
Class _____ Instructor _____

Learning Objective

● You will be able to identify the basic types of welding joints, the areas and parts of a completed weld, and the various welding positions.

Instructions

Carefully read Headings 3.1 through 3.4 of the text. Also study Figures 3-1 through 3-20 in the text. Then answer the following questions.

1. Name the five basic joints shown below.

1. A. _____
   B. _____
   C. _____
   D. _____
   E. _____

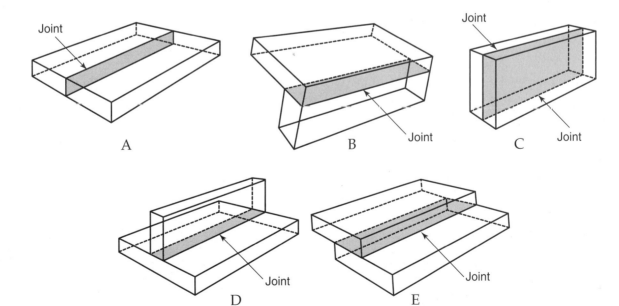

Copyright by The Goodheart-Willcox Co., Inc.   Modern Welding Lab Workbook   29

2. Label the areas and parts of the completed weld.

   A. _____
   B. _____
   C. _____
   D. _____
   E. _____
   F. _____
   G. _____

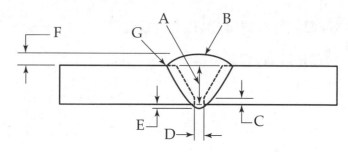

3. Name the various butt grooves shown.

   A. _____
   B. _____
   C. _____
   D. _____
   E. _____
   F. _____
   G. _____
   H. _____
   I. _____
   J. _____
   K. _____
   L. _____

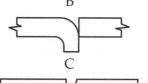
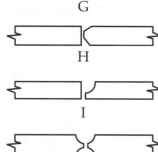
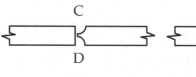
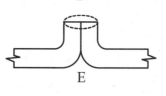
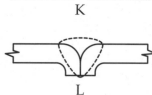
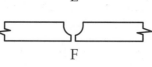

30   Modern Welding Lab Workbook   Copyright by The Goodheart-Willcox Co., Inc.

Lesson 3A   Welding Joints and Positions

Name _____

4. Name each type of corner joint shown.

   A. _____
   B. _____
   C. _____
   D. _____
   E. _____
   F. _____
   G. _____
   H. _____
   I. _____
   J. _____

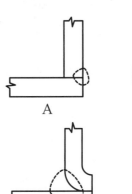
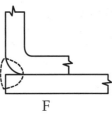
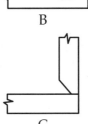
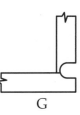

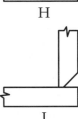
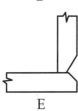
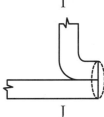
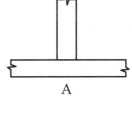
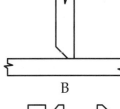

5. Identify each T-joint or edge joint.

   A. _____
   B. _____
   C. _____
   D. _____
   E. _____
   F. _____
   G. _____
   H. _____

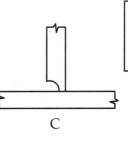
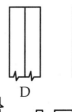
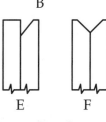

6. Weld positions are determined by the positions of the weld ____ and weld ____.

6. _____

7. The abbreviation for a groove joint in the overhead position is ____.

7. _____

8. The abbreviation for a fillet joint in the horizontal position is ____.

8. _____

9. The joint shown is a(n) ____.
    A. T-joint welded in the overhead position
    B. T-joint welded in the vertical position
    C. corner joint welded in the flat position
    D. corner joint welded in the horizontal position.
    E. edge joint welded in the flat position

9. _____

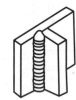

10. The joint shown is a(n) ____.
    A. T-joint welded in the overhead position
    B. T-joint welded in the vertical position
    C. corner joint welded in the flat position
    D. corner joint welded in the horizontal position
    E. edge joint welded in the flat position

10. _____

Lesson 3B

Reading Welding Symbols

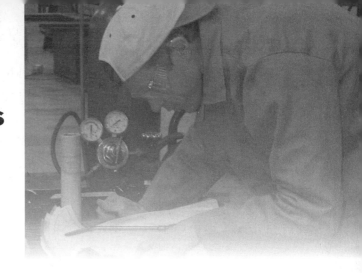

Name _____ Date _____
Class _____ Instructor _____

Learning Objective
- You will be able to describe the AWS welding symbol and demonstrate how it is used on a welding drawing.

Instructions
Carefully read Headings 3.5 through 3.6 of the text. Also study Figures 3-21 through 3-43 in the text. Then answer the following questions.

1. What is the difference between a welding symbol and a weld symbol?   1. _____
   A. The weld symbol gives all the information required to make a weld; the welding symbol does not.
   B. The weld symbol is part of the welding symbol.
   C. The welding symbol is part of the weld symbol.
   D. The weld symbol tells how strong the weld is to be.
   E. There is no difference.

2. Certain information may be given on a welding symbol. This information is always found in the same location on the welding symbol. Identify the information given by the letters or symbols indicated on the welding symbol shown.

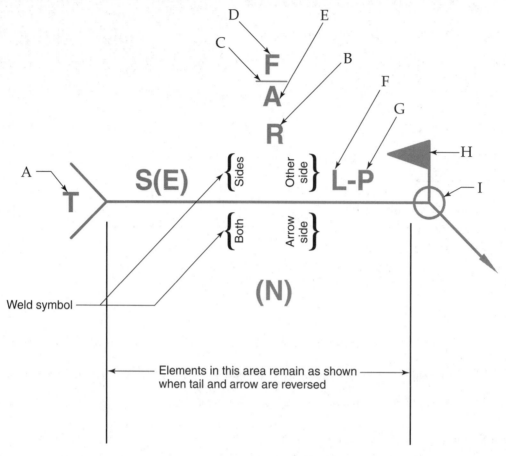

A. _____
B. _____
C. _____
D. _____
E. _____
F. _____
G. _____
H. _____
I. _____

3. *True or False?* A weld is to be made on the side of the joint that the arrow touches. The symbol for that weld is always shown below the reference line.

3. _____

4. *True or False?* The vertical line used in the bevel, fillet, and J-groove joint symbol is always drawn to the left side of the symbol drawing.

4. _____

Name _____

5. Given the information in the following drawings, first add the dimensions to the drawings in A through D. Then, sketch in the completed welds on the partial pieces to the right of the illustration shown.

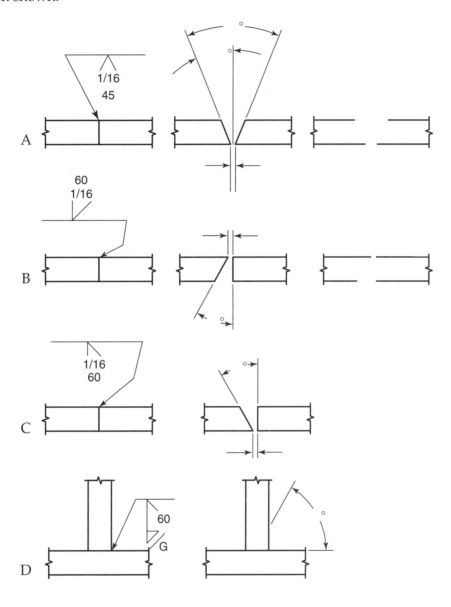

Lesson 3B   Reading Welding Symbols

6. For the following welding symbols, draw the two pieces of metal as they would be prepared prior to welding and dimension the angles of spaces required.

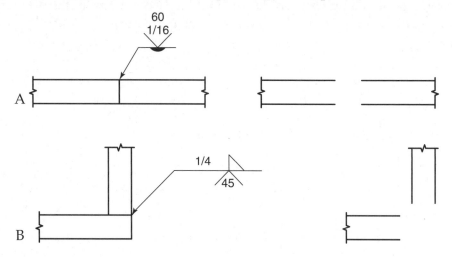

7. Show the shape of the weld face for the fillet welds made from these welding symbols.

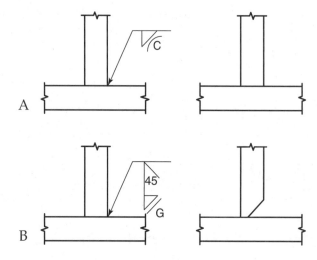

8. For the following weld finish symbols, what process is used to finish the surface of the finished weld?

C _____

G _____

M _____

9. The size of the weld and the effective throat size is always given to the _____ end of the _____ symbol.

9. _____

10. The size of a groove weld is the depth to which the base metal is _____.

10. _____

11. A weld that is made in repeated short lengths is called a(n) _____ weld.

11. _____

Name _____

12. Draw and dimension the intermittent weld indicated by the following symbol.

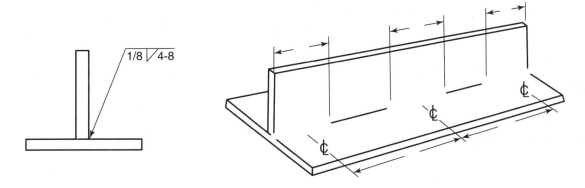

13. Draw and dimension the finished weld indicated by the following welding symbol.

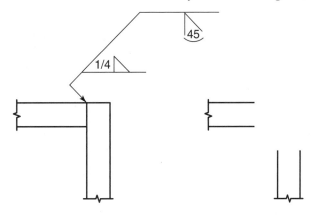

14. Draw the sides of the slot, the completed weld, and dimension the slot and weld that is required for the following welding symbol.

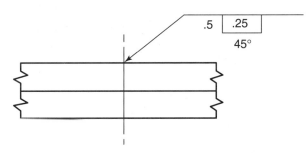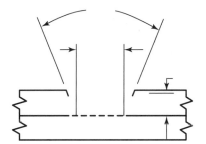

15. Answer the following questions about the weld symbols shown.

A. How wide is the weld in "A"?
B. What kind of a weld symbol is shown in "A"?
C. What welding process is used in "B"?
D. How many welds are made in "B"?
E. On what side of the metal is the weld in "B" made?

15. A. _____
    B. _____
    C. _____
    D. _____
    E. _____

Lesson 3C

Electrode Angles

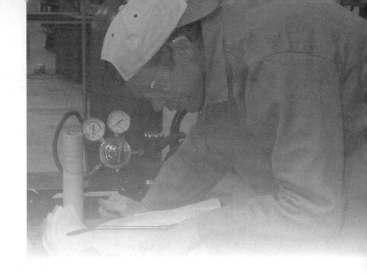

Name _____ Date _____
Class _____ Instructor _____

Learning Objective
You will be able to properly describe the correct position of an electrode, gun, or torch in relation to the material being welded.

Instructions
Carefully read Heading 3.7 of the text and its subheads. Also study Figures 3-44 through 3-48 in the text. Then answer the following questions.

1. The two terms used by the American Welding Society to describe the position of the electrode, gun, or torch in relation to the material being welded are _____ angle and _____ angle.

    1. _____

2. The _____ angle is the angle measured from a line perpendicular to the weld axis in the plane defined by the weld axis and the electrode axis.

    2. _____

3. When the welding end of an electrode points forward in the direction of travel, the angle is called a push travel angle. This is also known as _____ welding.

    3. _____

4. The _____ angle is the angle measured from a line perpendicular to the major or nonbutting surface to the plane containing the weld axis and the centerline of the electrode.

    4. _____

5. The work angle used for welding a butt joint is _____.
    A. 0°
    B. 45°
    C. 60°
    D. 90°

    5. _____

6. Drag or push travel angles usually range from zero to _____°.

    6. _____

7. A very large push angle, up to 85°, is used for _____.

    7. _____

Copyright by The Goodheart-Willcox Co., Inc.   Modern Welding Lab Workbook   39

8. The angle shown by the arrow in the following figure is a _____.
   A. work angle
   B. push travel angle
   C. drag travel angle

8. _____

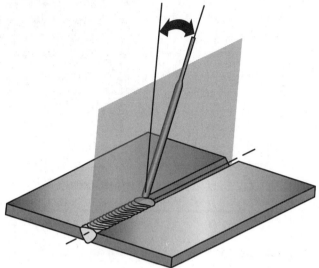

9. The angle shown by the arrow in the following figure is a _____.
   A. work angle
   B. push travel angle
   C. drag travel angle

9. _____

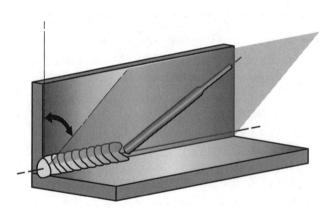

10. The angle shown by the arrow in the following figure is a _____.
    A. work angle
    B. push travel angle
    C. drag travel angle

10. _____

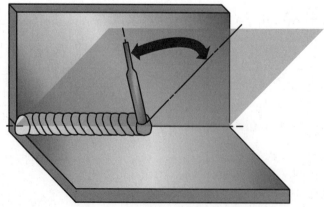

Lesson 4A

Welding and Cutting Processes

Name _____ Date _____

Class _____ Instructor _____

Learning Objective
- You will be able to identify and describe arc welding processes, oxyfuel joining processes, and oxygen cutting processes.

Instructions
Carefully read Headings 4.1 through 4.3 of the text. Also study Figures 4-1 through 4-18 in the text. Then answer the following questions.

1. Match the welding or cutting processes listed on the left with their accepted AWS abbreviations on the right.

   TS   A. Torch brazing   1. A. _____
   FCAW   B. Flux cored arc welding   B. _____
   TB
   OAW   C. Oxyacetylene welding   C. _____
   GMAW   D. Torch soldering   D. _____
   OFC
   GTAW   E. Gas tungsten arc welding   E. _____
   SMAW   F. Gas metal arc welding   F. _____
              G. Shielded metal arc welding   G. _____
              H. Oxyfuel gas cutting   H. _____

2. Name the parts indicated in the oxyacetylene welding outfit shown here.

   A. _____
   B. _____
   C. _____
   D. _____
   E. _____
   F. _____
   G. _____
   H. _____

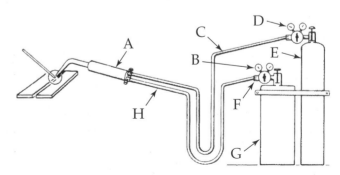

Copyright by The Goodheart-Willcox Co., Inc.

3. Give the temperatures of each item listed below in both °F and °C.
   A. A neutral oxyacetylene flame: _____°F (_____°C)
   B. Soldering is done below: _____°F (_____°C)
   C. Brazing is done above: _____°F (_____°C)
   D. An oxidizing oxy-LP gas flame:_____°F (_____°C)

3. A. _____
   B. _____
   C. _____
   D. _____

4. *True or False?* A consumable tungsten electrode is used with GTAW.

4. _____

5. The welding amperage in GTAW is determined by the _____.
   A. metal thickness
   B. electrode diameter
   C. type of metal welded
   D. All of the above.

5. _____

6. Name the items indicated in the SMAW station shown below.
   A. _____
   B. _____
   C. _____
   D. _____
   E. _____
   F. _____

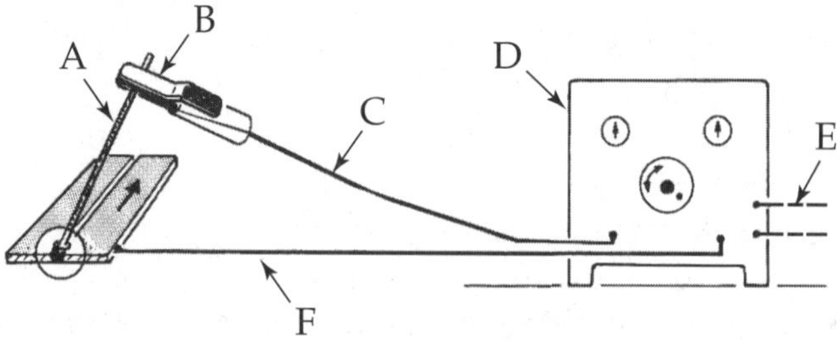

7. Explain why oxyfuel gas cutting is sometimes called "burning."
   _____
   _____
   _____

8. During OFC, what actually produces the kerf by causing the metal to burn and blow away rapidly?
   _____
   _____

9. List three pieces of safety equipment that must be worn when oxyacetylene welding or cutting.
   _____
   _____

10. A(n) _____ is used to clean the metal during TB.

10. _____

11. Why does TB cause less metal warping than is caused by OAW?
    _____
    _____

42   Modern Welding Lab Workbook

Lesson 4A Welding and Cutting Processes

Name _____

12. A severe health hazard may result due to toxic fumes that occur when metals containing _____ are brazed.
    A. zinc
    B. beryllium
    C. phosphorous
    D. cadmium
    E. All of the above.

12. _____

13. Name the parts indicated on the gas tungsten arc welding station in the following drawing.
    A. _____
    B. _____
    C. _____
    D. _____
    E. _____
    F. _____

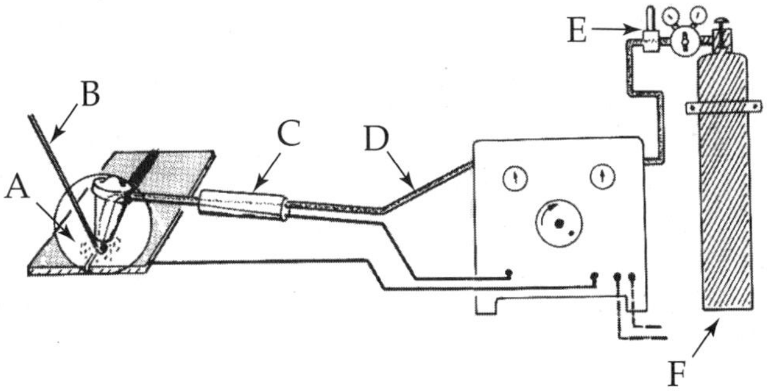

14. List three pieces of safety equipment that must be worn when SMAW, arc welding, or cutting.
    _____
    _____
    _____

15. GTAW torches may be _____ cooled or _____ cooled.

15. _____

16. Name the indicated parts on the GMAW outfit in the following illustration.
    A. _____
    B. _____
    C. _____
    D. _____
    E. _____
    F. _____
    G. _____
    H. _____
    I. _____
    J. _____
    K. _____
    L. _____

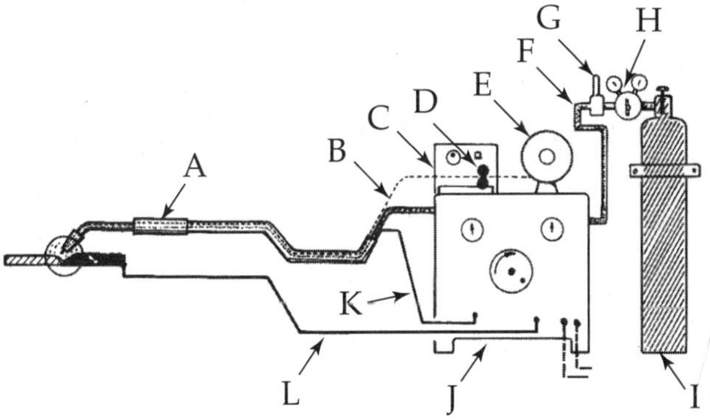

17. In FCAW, where is the flux material stored so that it is in constant use when welding?
    _____
    _____

Copyright by The Goodheart-Willcox Co., Inc.

Modern Welding Lab Workbook   43

18. In FCAW and GMAW, welding current is changed by changing the ____ feed speed.

18. _____

19. A hollow electrode is used in the ____ process.
   A. OC-P
   B. CAC-A
   C. OAC
   D. SMAC
   E. Both A and C.

19. _____

20. Oxygen from the oxygen cylinder is used for ____ when OAC.
   A. cutting the heated metal
   B. cooling the heated metal
   C. cooling the electrode
   D. cooling the welder's hand
   E. cooling the electrode holder

20. _____

21. *True or False?* An oxygen lance is burned up in the lance cutting process.

21. _____

22. During underwater cutting, the oxyfuel cutting flame is kept from going out by the ____.
   A. pressure of the cutting oxygen
   B. seawater
   C. air pocket around the flame
   D. movement of the torches

22. _____

23. Flux powder reduces the formation of ____ and makes cutting easier when OC-F.

23. _____

24. *True or False?* Exothermic cutting is also known as ultrathermic cutting.

24. _____

25. In metal powder cutting, ____ powder is added to the flame instead of flux.

25. _____

Lesson 4B

Welding and Cutting Processes

Name _____ Date _____

Class _____ Instructor _____

Learning Objective
- You will be able to identify and describe resistance welding processes, arc cutting processes, and specialized arc welding processes.

Instructions
Carefully read Headings 4.4 through 4.6 of the text. Also study Figures 4-19 through 4-28 in the text. Then answer the following questions.

1. The AWS abbreviation for resistance spot welding is ____.
   A. RSW
   B. RSEW
   C. RW
   D. SW

   1. _____

2. The AWS abbreviation for air carbon arc cutting is ____.

   2. _____

3. *True or False?* The AWS abbreviation for electroslag welding is EW.

   3. _____

4. A(n) ____ joint is used in resistance spot welding to weld together two or more pieces.

   4. _____

5. When resistance spot welding, the operator must wear ____ and ____.

   5. _____

Copyright by The Goodheart-Willcox Co., Inc.   Modern Welding Lab Workbook   45

6. Name the parts indicated in the following plasma arc cutting illustration.

   A. _____
   B. _____
   C. _____
   D. _____
   E. _____
   F. _____
   G. _____
   H. _____

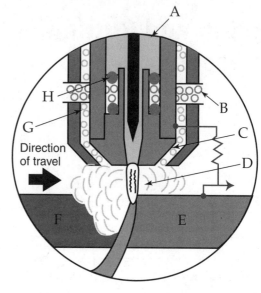

7. Describe what is meant by "projections" in the projection welding process.

   _____

   _____

8. Name the parts indicated on the PW schematic.

   A. _____
   B. _____
   C. _____
   D. _____
   E. _____
   F. _____
   G. _____
   H. _____

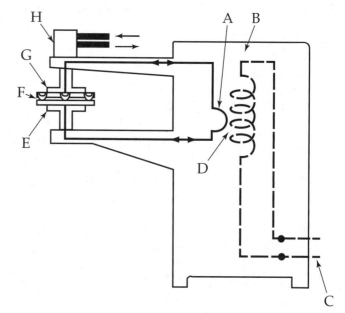

9. Resistance seam welding uses electrodes in the form of _____.   9. _____

10. Which of the following is *not* true of flash welding?   10. _____
    A. The finished weld has an enlargement at the joint.
    B. An electric arc heats the base metals.
    C. Light pressure is used to fuse the heated ends of the two metals.
    D. Flying sparks are produced during the process.
    E. The process creates a strong, clean weld joint.

Name _____

11. Name the areas of the flash weld shown.

    A. _____
    B. _____
    C. _____

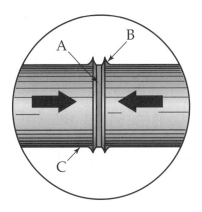

12. *True or False?* The PAC process can be used to cut nonmetals.   12. _____

13. What causes the base metal to be blown away from the heated area during the PAC process?
    _____
    _____

14. The electrode used when SMAC is heavily covered with _____.   14. _____

15. Name the parts indicated in the CAC-A drawing.

    A. _____
    B. _____
    C. _____
    D. _____
    E. _____
    F. _____

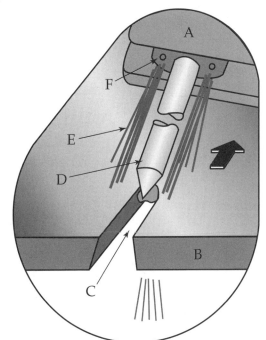

16. *True or False?* The air used in the air carbon arc cutting process may be supplied from a compressed air cylinder.

16. _____

17. Submerged arc welding is called "submerged" because the welding is done under _____.
    A. water
    B. a heavy gas shield
    C. heavy metal
    D. a layer of molten metal
    E. a heavy layer of flux

17. _____

18. *True or False?* Electroslag welding is used to weld joints on very thin metal.

18. _____

19. The molten metal is kept from running out of the weld in ESW as the weld moves up by sliding copper _____.

19. _____

20. Which welding process is generally used to attach bolts, screws, rivets, and spikes to metal surfaces?

20. _____

Lesson 4C

Welding and Cutting Processes

Name _____ Date _____

Class _____ Instructor _____

Learning Objective
● You will be able to identify and describe solid-state welding processes and other welding processes, including laser beam, electron beam, and torch plastic welding. You will also be able to describe the relationship between steel temperature and color.

Instructions
Carefully read Headings 4.7 through 4.9 of the text. Also study Figures 4-29 through 4-38 in the text. Then answer the following questions.

1. In solid-state welding processes, pressure is applied at a welding temperature below the _____ temperature of the base metal.

   1. _____

2. Even when cold welding, the welder should wear _____
   _____

3. Name the parts of the CW process.
   A. _____
   B. _____
   C. _____
   D. _____

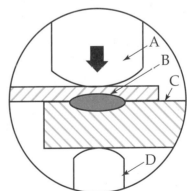

Copyright by The Goodheart-Willcox Co., Inc.   Modern Welding Lab Workbook   49

4. Which welding process uses a tremendous shock wave to fuse (weld) the pieces together?
   _____

5. Which of the following is true of the FOW process?   5. _____
   A. Friction is used to fuse two pieces of metal together.
   B. Hot coke is used to heat the parts to be welded.
   C. A beam of energy is used to heat the metals being joined.
   D. No outside heat source is used.

6. In FRW, how is enough friction heat created to cause steel to melt and weld?
   _____
   _____

7. A friction stir welding tool is ____ at high speed to create a   7. _____
   weld.

8. How is the degree of heating controlled during ultrasonic welding?
   _____
   _____
   _____

9. The ultrasonic welding process is used most often to join   9. _____
   ____ materials.
   A. very light materials
   B. very heavy materials
   C. thick materials
   D. plastics

10. *True or False?* Mirrors are used to direct laser welding beams   10. _____
    to the weld joint.

11. *True or False?* The electron beam welder produces welds on   11. _____
    thick metal that are deep and wide.

12. Electron beam welding uses the energy of a focused stream of   12. _____
    ____ to heat and weld metals.

13. Plastic is generally welded at temperatures of ____°F   13. _____
    (____°C) to____°F (____°C) by heated air or gas.

14. The color of steel as it is heated indicates the temperature of the metal fairly accurately. List the
    following temperatures:
    Bright yellow: _____°F (_____°C)
    Orange red: _____°F (_____°C)
    Deep straw: _____°F (_____°C)

15. What color is the surface of steel at 1000°F (538°C)?   15. _____

50   Modern Welding Lab Workbook

Lesson 5A
SMAW Equipment and Supplies

Name _____ Date _____

Class _____ Instructor _____

Learning Objective
You will be able to describe the various types of shielded metal arc welding (SMAW) machines. You will be able to determine the various accessories and supplies required in SMAW. You will also be able to select the correct current.

Instructions
Carefully read Headings 5.1 through 5.3.1 of the text. Also study Figures 5-1 through 5-30 in the text. Then answer the following questions.

1. A complete SMAW station consists of a(n) _____.
   A. AC or DC arc welding machine
   B. electrode holder
   C. electrode lead
   D. workpiece lead
   E. Both A and B.
   F. All of the above.

   1. _____

2. Which of the following is *not* true of an inverter power supply compared to a regular transformer-type power supply?
   A. More efficient.
   B. Lighter weight.
   C. Larger transformer.
   D. Smaller size.

   2. _____

3. *True or False?* The general slope of the volt-ampere curve is called the output slope of the power source.

   3. _____

Modern Welding Lab Workbook   51

4. Which of the following describes the volt-ampere curve shown?
   A. Constant current
   B. Steep slope
   C. Constant voltage
   D. Both A and B.
   E. Both B and C.

4. _____

5. Droopers are _____ arc welding machines.
   A. AC
   B. DC
   C. constant current
   D. constant voltage
   E. AC/DC

5. _____

6. The constant _____ machine is most desirable when doing manual arc welding.

6. _____

7. *True or False?* A manual arc welder would normally prefer a flatter constant current output slope when welding in a horizontal, vertical, or overhead position.

7. _____

8. List the three principal components of the transformer type arc welder.

_____

_____

_____

Lesson 5A SMAW Equipment and Supplies

Name _____

9. What type of control is shown in the schematic?

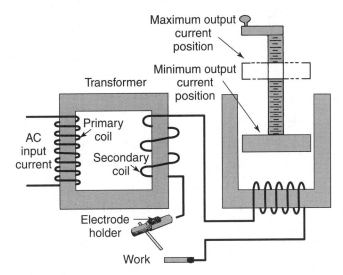

_____

10. The symbol for a device that changes AC to DC by allowing current to flow only in one direction is shown below. What is this device called?

10. _____

11. What do the letters DCEN mean?

_____

12. DCSP has the same direction of current flow as DCE _____.

12. _____

13. If a welder varies the arc gap (voltage) of the system and only a very small change in the delivered current occurs, the welding machine is called a constant _____ arc welding machine.

13. _____

14. What do the letters DCEP mean?

_____

15. What is the rated load (welding) voltage for a Class I, 400 ampere arc welding machine? Show your calculations below:

15. _____

16. Describe the term *duty cycle*.

   _____

   _____

   _____

17. Match each of the actions in an inverter power supply's operation with its proper order.

   A. Step 1   _____ Incoming AC current is rectified to DC current.
   B. Step 2   _____ DC current is smoothed using an inductor.
   C. Step 3   _____ High-frequency current is passed through the transformer.
   D. Step 4   _____ DC current is passed through the inverter to create high-frequency AC current.
   E. Step 5
                   _____ Output rectifier changes current to DC

18. A duty cycle-welding amperage chart is shown in Figure 5-30 of the text. How many minutes out of every 10 minutes can this machine operate continuously at 125 amperes?

    18. _____

19. What NEMA arc welding machine classification is assigned to a machine that can deliver its rated output at a constant duty cycle of 30%–50%?

    19. _____

20. What duty cycle is recommended for automatic and semiautomatic arc welding operations?
    A. 20
    B. 40
    C. 60
    D. 80
    E. 100

    20. _____

54   Modern Welding Lab Workbook

Lesson 5B SMAW Equipment and Supplies

Name _____

15. _____ more filler metal is deposited with an E7028 iron powder electrode than is deposited with the same diameter E6012 electrode.
    A. 0%
    B. 22%
    C. 45%
    D. 78%
    E. 100%

15. _____

16. What polarity is used with a chromium or chromium-nickel steel electrode with the number **E310-16**?

16. _____

17. Which AWS electrode specification contains specifications for low-alloy SMAW steel electrodes?

17. _____

18. To eliminate the time required to travel back and forth to the welding machine, several manufacturers provide _____ devices which may be kept near the operator for convenient control of the machine.

18. _____

19. Name the item shown in the following image.

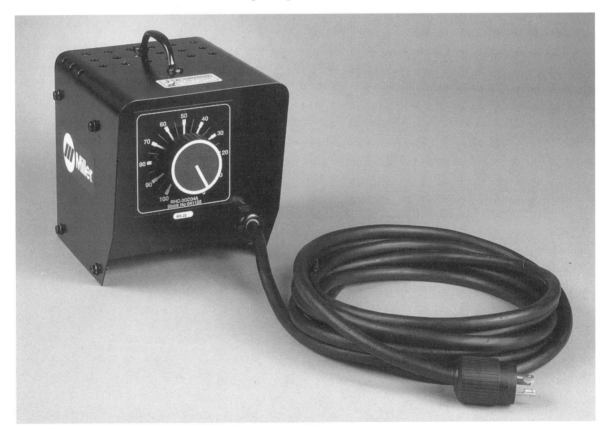

_____

20. Identify the item shown below and describe its function.

_____
_____
_____
_____

21. Lens shade # _____ is suggested for use in shielded metal arc welding on 1/4″ (6.4mm) metal.
    A. 10
    B. 11
    C. 12
    D. 13
    E. 14

21. _____

Lesson 6A

Shielded Metal Arc Welding Safety

Name _____ Date _____

Class _____ Instructor _____

> **Learning Objective**
> • You will be able to use safe work practices when shielded metal arc welding (SMAW).
>
> **Instructions**
> Carefully read Chapter 5 and Chapter 6 of the text. Then answer the following questions.

1. The line voltage to an industrial arc welding machine is generally 220V or ____V. Because of this, only an experienced electrician should work on the electrical power connections used in an arc welding machine.

   1. _____

2. A # ____ filter lens should be worn when doing SMAW on 1/4" (6.4mm) metal for long periods of time. Also, a pair of ____ goggles should be worn to reduce eye damage from flashes behind the helmet.

   2. _____

3. The inlet to the ventilation pickup duct should be located so that fumes are removed before they reach the welder's ____.

   3. _____

4. Which of the following must be clean and tight to reduce electrical resistance? (More than one answer may apply).
   A. Booth curtains.
   B. Electrode holder jaws.
   C. Workpiece lead at the work.
   D. Workpiece lead at the electrode holder.
   E. Electrode lead at the machine and holder.

   4. _____

Copyright by The Goodheart-Willcox Co., Inc.

Modern Welding Lab Workbook   59

5. Name six hazards to avoid during arc welding.
_____
_____
_____
_____

6. *True or False?* You should avoid using arc welding equipment in damp places or with damp welding gloves.

6. _____

7. Select the one statement below that is *not* important when working around arc welding equipment.
   A. Never look at the arc from any distance without wearing an approved filter lens.
   B. All pockets should be covered.
   C. Do not carry plastic pens, combs, or matches in your pockets.
   D. Never work without adequate ventilation.
   E. Take every precaution to eliminate H_2O and O_2.

7. _____

8. An arc welding machine must never be started under load. Because of this, the electrode holder must always be hung on a(n) _____ hanger before the machine is started.

8. _____

9. List six items that a welder should wear when arc welding in the overhead position.
_____
_____
_____
_____

10. Why should homemade or non-NEMA approved transformer equipment never be used?
_____
_____
_____

Lesson 6B

Shielded Metal Arc Welding Fundamentals

Name _____ Date _____

Class _____ Instructor _____

Learning Objective
- You will be able to describe the fundamentals of the shielded metal arc welding (SMAW) process. You will also be able to strike an arc, run a bead, and "read" a bead.

Instructions
Carefully read Headings 6.1 through 6.6.5 of the text. Also study Figures 6-1 through 6-25 in the text. Then answer the following questions.

1. The temperature of the arc in SMAW is ____°F (____°C) to ____°F (____°C).

    1. _____

2. What is the polarity of the following welding circuit?

    2. _____

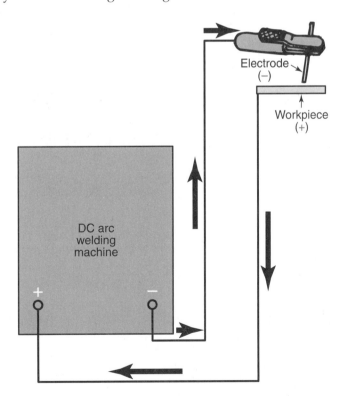

3. In a constant current machine, if the arc gap is increased, the resistance will _____ and the voltage output of the machine must _____.

3. _____

4. The decision to use DCEN or DCEP often depends on such variables as the metal thickness, the position of the joint, the depth of penetration desired, the type of base metal, and _____.

5. Electrodes intended for use with AC have _____ agents in the covering to help _____ the arc.

5. _____

6. Which of the following is *not* an advantage of an AC SMAW welding machine?
   A. AC welding machines are less expensive than DC welding machines.
   B. All SMAW electrodes can be used with AC.
   C. Large-diameter electrodes can be used with high AC currents.
   D. Penetration is moderate.

6. _____

7. Name the various parts of the arc weld and electrode shown in the following drawing.
   A. _____
   B. _____
   C. _____
   D. _____
   E. _____
   F. _____
   G. _____
   H. _____

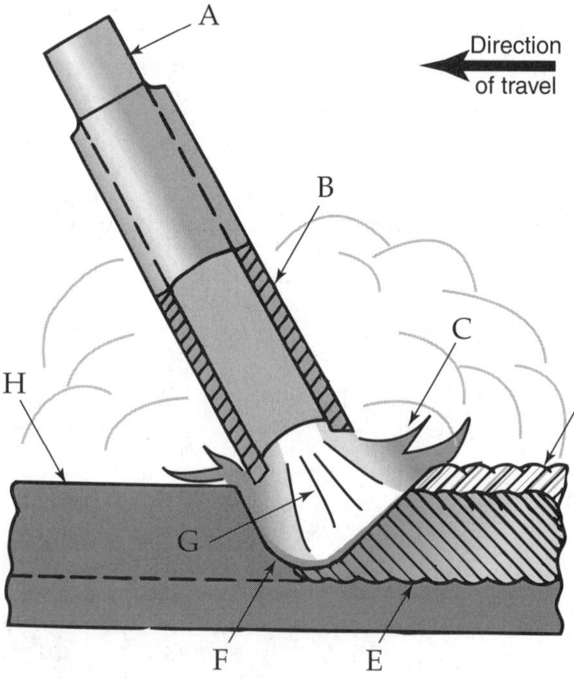

8. Before a thorough inspection of the arc welding station can be made, the arc welding machine should be turned _____ to check it properly.

8. _____

9. Assuming a 5/32" (4mm) diameter electrode is used, answer the following questions.
   A. What is the maximum width of a stringer bead?
   B. What is the maximum width of a weave bead?
   C. What is the travel angle used?

9. A. _____
   B. _____
   C. _____

Name _____

10. Name the two methods for striking (starting) a shielded metal arc shown in the following drawing.

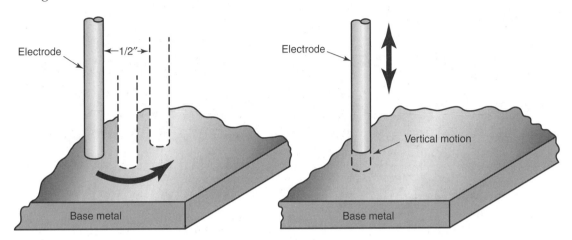

A. _____

B. _____

11. List five factors that a welder must control or select in order to produce a good weld using the SMAW process.

_____
_____
_____
_____
_____

12. List the four factors that an arc welder must control in order to produce an acceptable bead.

_____
_____
_____
_____

13. Much can be learned by watching the arc bead as it is being made and when it is completed. From the appearance of the following beads, describe the conditions that caused the bead shapes shown.

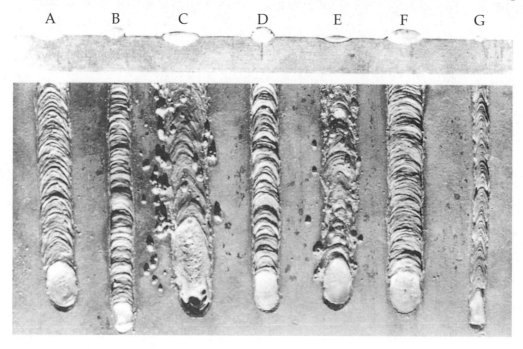

A. _____
B. _____
C. _____
D. _____
E. _____
F. _____
G. _____

14. The following illustration shows the steps to restarting an arc bead. Explain what must be done in each of the three steps.

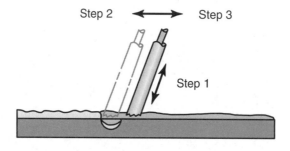

Step 1: _____
Step 2: _____
Step 3: _____

15. The slag coating must be removed before a second bead is placed on top of the first bead. This procedure should prevent slag ____.

15. _____

Job 6B-1

Inspecting and Setting Up an Arc Welding Station

Name _____ Date _____

Class _____ Instructor _____

Learning Objective

- In this job, you will demonstrate your ability to thoroughly inspect an SMAW, GMAW, FCAW, GTAW, or a PAC station, as assigned by the instructor.

1. Make certain the machine is turned off. Proceed with your inspection. Place an (X) in the "OK" column if the item checked is in good condition. Place an (X) in the "Needs Repair" column if the item needs repair.

	OK	Needs Repair

2. Check the work lead and electrode lead gas and water hoses for:

   A. Tight connections at the machine.

   B. Tight connections at the ground clamp or workpiece lug.

   C. Tight connections at the electrode holder or gun.

   D. Cuts, worn areas, or deep cracks over the full length of each lead.

   E. Tight water and gas hose connections.

   F. Cuts, worn areas, or deep cracks over the full length of each hose.

3. Make certain that a fully insulated electrode hanger is located near the worktable.

4. Turn on the ventilation system. Hold a large piece of paper near the ventilation system intake to determine that the ventilation system is pulling air from the booth.

5. Check the walls of the booth and curtains for holes that may need repair.

6. Ensure that the pliers, chipping hammer, and wire brush you may use are in good working condition.

7. Check your welding helmet for holes or cracks.

8. Check that your filter lens is the correct number and that the cover lens is clean and in good condition.

Instructor's initials: _____

Job 6B-2

Striking an Arc and Running Short Beads

Name _____ Date _____
Class _____ Instructor _____

Learning Objective
- In this job, you will practice striking the arc and running short beads.

1. Obtain a piece of mild steel that measures 1/4" × 6" × 6" (6.4mm × 150mm × 150mm) and four 1/8" (3.2mm) diameter, E6012 electrodes to begin this job.

   Note: This plate will be used for several jobs, so do not lose it.

2. Using a rule and a piece of soapstone or chalk, draw lines 3/8" (9.5mm) apart across the full 6" (150mm) length of the metal on both sides. See the following diagram.

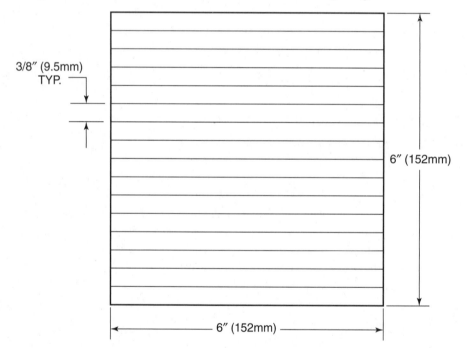

3. Answer the following questions. Set your machine and have the instructor check your settings.
   A. What type electrode covering does this electrode have?   3. A. _____
   B. With what type current can it be used?   B. _____
   C. In what position will you be welding?   C. _____
   D. What range of current should you use for this electrode?   D. _____
      (**Note:** Set the machine to the middle of this range.)
   E. What is the tensile strength of this electrode?   E. _____
                                                                          _____

66   Modern Welding Lab Workbook   Copyright by The Goodheart-Willcox Co., Inc.

Job 6B-2　Striking an Arc and Running Short Beads

Name _____

4. Button your shirt or coveralls at the neck, put your welding helmet on and tighten the headband, put on your welding gloves, and prepare to weld.
5. Your metal should be placed on the table so that the 6″ (150mm) length is running from your left to right. See previous figure.
6. Lower your arc helmet and attempt to strike the arc. Try both the straight down-and-up motion or the scratching motion. Once the arc is struck, carefully adjust the arc length to a distance approximately equal to the diameter of the electrode. Once the arc is stabilized, move forward in the left-to-right direction slowly for about 1″ (25mm) and withdraw the electrode so the arc is stopped. (**Note:** If you are left-handed, you should weld in a right-to-left direction.)

   ▪ **Caution:** Beginners sometimes cause the electrode to weld itself to the base metal when the arc is struck. If this occurs, proceed as follows:

   A. With your arc helmet still down, release the electrode from the electrode holder.
   B. Raise your arc helmet.
   C. Hang the electrode holder on the insulated hanger.
   D. Bend the electrode back and forth until it breaks away from the base metal.
   E. Reinstall the electrode in the electrode holder.
   F. Lower your arc helmet and try again.

7. If the electrode sticks a great deal, increase the current setting on the machine about 10A and try again. If the electrode spatters a great deal, try decreasing your arc gap or decrease the current setting on the machine about 10A.
8. Try to restrike the arc just ahead of the first attempt and run another 1″ (25mm) bead. Then stop the arc.
9. Continue striking and stopping the arc until the top two lines on your plate are covered with 1″ (25mm) long beads.

Instructor's initials: _____

Job 6B-3

Running Arc Beads

Name _____ Date _____

Class _____ Instructor _____

> **Learning Objective**
> - In this job, you will be running (creating) several arc beads.

1. Use the same side of the plate you were given to practice striking the arc in the previous job. Obtain six 1/8" (3.2mm) diameter, E6012 electrodes to begin this job.
2. Set the machine polarity. Set the amperage at the mid-range for the electrode being used.
3. Lower your arc helmet. Strike the arc, and run a stringer bead (no sideways electrode motion) continuously along the fourth marked line.
4. The bead should be about three times as wide as the electrode diameter. The ripple shape at the rear of the weld pool should be bullet-shaped if the forward speed is correct. The electrode should be changed when the total length of the electrode gets to be about 2 1/2" (65mm) or otherwise as specified by your instructor. Place a new electrode into the holder and restart the bead as described in Heading 6.6.4 of the text.

   ■ **Note:** As you weld, watch the width of the weld pool and bead. Also, to ensure a proper speed, watch the rear of the bead. It should always appear bullet-shaped. If the bead width or the rear of the bead do not look right, change your forward speed, electrode angle, or machine settings until bead appearance is correct.

5. Wear chipping goggles whenever you are cleaning your beads. Chip and wire brush your bead until the slag coating is removed.
6. Check the appearance of your first bead against those shown in Figure 6-23 in the text. If the bead is not correct, change what needs to be changed before you run the second bead.
7. Starting with your sixth marked line, run five additional beads, leaving one marked line between beads.
8. Chip and clean each bead prior to beginning the next bead. Make corrections to the machine settings, your electrode angle, travel speed, arc gap, bead width, and bead shape as required after you inspect and evaluate each bead.

> **Inspection**
> The beads should be straight, about 3/8" (9.5mm) wide. They must have a uniform buildup, evenly spaced ripples, and good fusion with no overlap or undercutting at the toes.
>
> Instructor's initials: _____

Job 6B-4

Running Weave Beads

Name _____ Date _____

Class _____ Instructor _____

Learning Objective
- You will run several weave beads using an E6012 electrode.

1. Obtain about six 1/8″ (3.2mm) diameter, E6012 electrodes to begin this job.
2. Using a rule and a piece of soapstone or chalk, draw lines 3/8″ (9.5mm) apart across the full 6″ (150mm) length of the metal on the reverse side of your original piece. See the following figure.

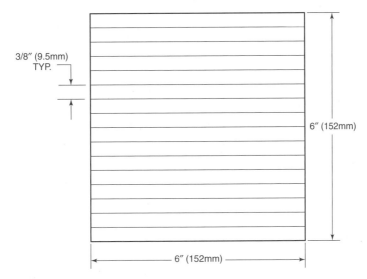

3. Set the machine amperage and polarity.
4. For this job, use the side opposite the one used for the previous two jobs.
5. The second line from the top should be used as the center point of your first bead. Make a weave bead approximately 3/4″ (20mm) wide or six times the electrode diameter.
6. Chip and wire brush this bead.
7. Read the bead. Inspect it, and decide what changes need to be made to improve its appearance and quality.
8. Use the 4th, 6th, and 8th lines as centers and make three additional weave beads.

   ■ **Note:** Constantly "read the bead" and make needed changes in speed and electrode angle as you weld.

Inspection
The beads should be straight, about 3/4″ (20mm) wide. They must have a uniform buildup, evenly spaced ripples, and good fusion with no overlap or undercutting at the toes.

Instructor's initials: _____

Job 6B-5

Running Stringer Beads

Name _____ Date _____

Class _____ Instructor _____

> **Learning Objective**
> - You will run stringer beads with E6010, E6011, E6013, and other electrodes as selected by the instructor.

1. Obtain a second piece of mild steel that measures 1/4" × 6" × 6" (6.4mm × 150mm × 150mm) and mark it with lines 3/8" (9.5mm) apart, using a soapstone, as you did in job 6B-2.
2. Using 1/8" (3.2mm) E6010 electrodes, make five stringer beads using drawn lines #2, 4, 6, 8, and 10 as the centers for the beads.
3. Answer the following questions before beginning to weld. (See Figures 6-15 and 6-17 in the text.)
   A. What amperage range is suggested?   3. A. _____
   B. What type of current is required?   B. _____
   C. What type of covering is on this electrode?   C. _____
4. Clean, inspect, and read each bead before making the next bead. Change any setting or other variable required to make the next bead better.
5. Using 5/32" (4mm) E6011 electrodes, make four weave beads on the side opposite the weave beads made in Step 2. Answer the following questions before beginning to weld.
   A. What amperage range is suggested?   5. A. _____
   B. What type(s) of current can be used?   B. _____
   C. What type of covering does the electrode have? C. _____
6. Read each bead and make corrections before making the next bead.
7. Using 3.32" (2.4mm) E6013 electrodes, make four stringer beads next to each other. Weld wherever space is available, or on top of previous beads. Answer the following questions before beginning to weld.
   A. What amperage range is suggested?   7. A. _____
   B. What type of current may be used?   B. _____
   C. What type of covering is on this electrode?   C. _____
8. Clean and read each bead and before making the next bead.

> **Inspection**
>
> All beads should be even in width, straight, with uniform buildup and ripple spacing. No overlap or undercutting should be present at the toes of the weld. The stringer bead width should be three times the electrode diameter. Weave beads should be six times the electrode diameter.

Instructor's initials: _____

70   Modern Welding Lab Workbook

Lesson 6C

SMAW in the Flat Welding Position

Name _____ Date _____

Class _____ Instructor _____

Learning Objective
- You will be able to perform SMAW in the flat welding position.

Instructions
Carefully read Headings 6.7 through 6.9.4 of the text. Also study Figures 6-26 through 6-48 in the text. Then answer the following questions.

1. In AC, the _____ direction of current virtually cancels the magnetic blow effect.

   1. _____

2. During welding using DC, the arc has a tendency to _____ as a weld nears the end of the joint.
   A. blow toward the end of the joint
   B. blow toward the beginning of the joint
   C. remain steady
   D. blow to the right side
   E. blow to the left side

   2. _____

3. What type of weld is described below?

   The weld axis is horizontal, and the weld face is horizontal or near-horizontal. _____

4. Name the parts of a fillet weld shown in the following drawing.

   A. _____
   B. _____
   C. _____
   D. _____
   E. _____
   F. _____

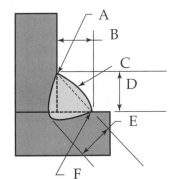

Copyright by The Goodheart-Willcox Co., Inc.   Modern Welding Lab Workbook   71

5. Name the parts of the groove weld shown in the following drawing.

   A. _____
   B. _____
   C. _____
   D. _____
   E. _____

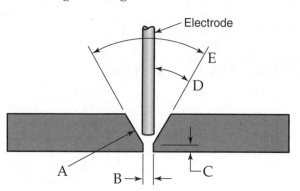

6. What type of defect is shown in the following welds?

   A. _____
   B. _____

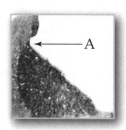
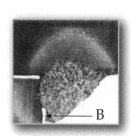

7. How wide should the bead generally be when making an edge weld?

   _____
   _____

8. Name the parts of a groove weld in progress shown in the following drawing.

   A. _____
   B. _____
   C. _____

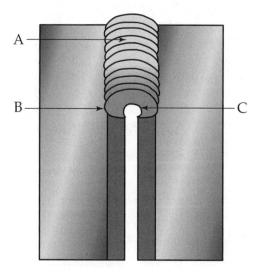

Name _____

Lesson 6C SMAW in the Flat Welding Position

9. A. How many beads were used to make the groove weld shown in the following drawing?
   B. What is the proper name of the first bead?

9. A. _____

   B. _____
   _____

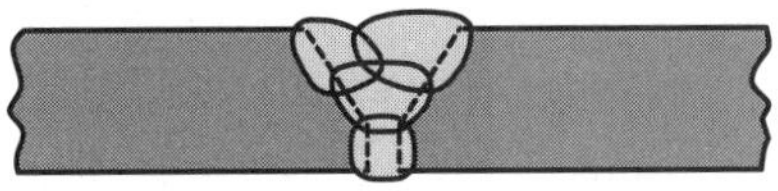

10. List three corrective measures that may be taken to prevent or reduce arc blow.

_____
_____
_____
_____
_____
_____
_____
_____
_____
_____
_____
_____
_____
_____
_____
_____

Job 6C-1

Square-Groove Weld on an Edge Joint in the Flat Welding Position

Name _____ Date _____

Class _____ Instructor _____

Learning Objective

- In this job, you will learn to make a square-groove weld on an edge joint in the flat welding position.

1. Obtain two pieces of mild steel that measure 1/4″ × 5″ × 5″ (6.4mm × 125mm × 125mm).
2. Obtain five E6010 electrodes of the smallest suggested diameter.
3. Refer to Figures 6-15 and 6-17 in the text, and answer the following questions for the E6010 electrode used:

    A. What diameter should be used?
    B. What current type and polarity should be used?
    C. What current range should be used?

    3. A. _____
    B. _____
    C. _____

4. What number welding filter lens should be used? (Refer to Figure 5-57 in the text.)

    4. _____

5. Place the pieces together and tack weld the weldment twice on each side.
6. Make an edge weld in the flat welding position on one edge. Make an edge weld on each of the four sides as shown in the following drawing. Reposition the pieces after welding each edge so all welds are made in the flat position.

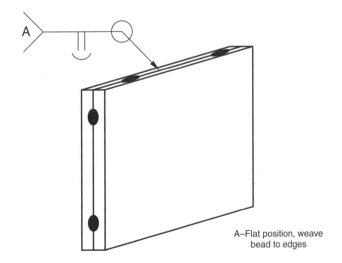

A—Flat position, weave bead to edges

Inspection

Your beads should be complete, with no voids or low spots where the beads stop and start. The beads should have evenly spaced ripples, with an even buildup. There should be no overlapping at the metal edges.

Instructor's initials: _____

Job 6C-2

Fillet Weld on a Lap Joint in the Flat Welding Position

Name _____ Date _____
Class _____ Instructor _____

> **Learning Objective**
>
> • In this job, you will learn to produce a fillet weld on a lap joint in the flat welding position.

1. Obtain four pieces of mild steel that measure 1/4" × 1 1/2" × 5" (6.4mm × 40mm × 125mm).
2. Obtain five E6011 electrodes of the largest suggested diameter.
3. Refer to Figures 6-15 and 6-17 of the text and answer the questions below for the electrode and diameter used.

   A. What diameter electrode should be used? 3. A. _____
   B. What type and polarity or current should be used? B. _____
   C. What current range should be used? C. _____

4. What number welding lens filter should be used? (Refer to Figure 5-57 in the text.) 4. _____

5. Place the pieces together, and tack weld them twice on each side. Place the pieces so the fillet weld will be made in the flat position.
6. Make two fillet welds as shown in the following drawing.

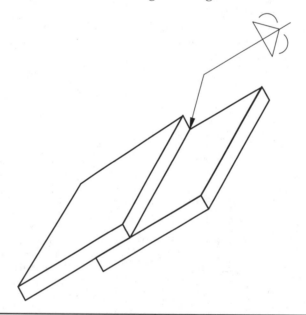

> **Inspection**
>
> Your fillet welds should be straight and even in width. The ripples in the beads should be evenly spaced, and the beads should be convex. There should be no undercutting or overlapping.
>
> Instructor's initials: _____

76   Modern Welding Lab Workbook

Job 6C-3

Fillet Weld on a T-Joint in the Flat Welding Position

Name _____ Date _____

Class _____ Instructor _____

Learning Objective

• In this job, you will make a fillet weld on a T-joint in the flat welding position.

1. For this job, you will need two pieces of mild steel that measure 3/16″ × 1 1/2″ × 5″ (4.8mm × 40mm × 125mm).
2. Obtain five E6012 electrodes of the proper size.
3. Refer to Figures 6-15 and 6-17 in the text, and answer the following questions.
   A. What diameter electrode should be used? 3. A. _____
   B. What type and polarity or current should be used? B. _____
   C. What current range should be used with this electrode? C. _____
4. What number filter lens should be used? (Refer to Figure 5-57 in the text.) 4. _____
5. Place the pieces into the proper position and tack weld them on each side of the joint about 3″ (75mm) apart. Place the tack welded pieces so the weld will be made in the flat position.
6. Make the welds as shown in the following drawing.

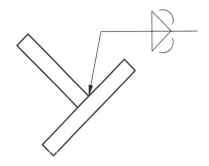

Inspection

Your fillet welds should be straight, with an even width, and evenly spaced ripples. They should be convex in shape. They should be free from undercutting or overlapping.

Instructor's initials: _____

Job 6C-4

Bevel-Groove Weld on a Butt Joint in the Flat Welding Position

Name _____ Date _____

Class _____ Instructor _____

> **Learning Objective**
>
> • In this job, you will learn to make a bevel-groove weld on a butt joint in the flat welding position.

1. Obtain four pieces of mild steel that measure 1/4" × 1 1/2" × 5" (6.4mm × 40mm × 125mm).
2. You will need five E6013 electrodes for this job.
3. Refer to Figures 6-15 and 6-17 in the text, and answer the following questions:
   A. What diameter electrode should be used? _____
   B. What type and polarity or current should be used? _____
   C. What current range should be used? _____

4. What number welding filter lens should be used?   4. _____
5. Prepare the bevel. Place the pieces in their proper position and tack weld each joint.
6. Make the weld as shown in the following drawing. Make certain that the welds are made on the correct side of the metal.

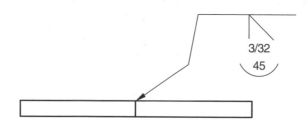

> **Inspection**
>
> Your welds should be straight, even in width, with all beads having evenly spaced ripples. There must be 100% penetration, with the penetration showing on the root side of each joint.
>
> Instructor's initials: _____

78   Modern Welding Lab Workbook

Lesson 6D

SMAW in the Horizontal Position

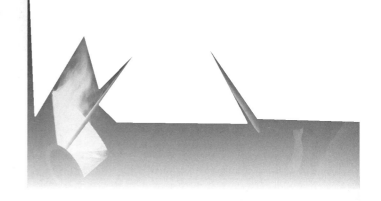

Name _____ Date _____
Class _____ Instructor _____

Learning Objective
- You will be able to perform SMAW in the horizontal position.

Instructions
Carefully read Heading 6.9.5 of the text. Also study Figures 6-49 through 6-52 in the text. Then answer the following questions.

1. Why is the electrode pointed upward at about 20° when performing a horizontal weld?
   _____
   _____

2. Gravity causes the metal in the air to fall as it travels across   2. _____
   the arc from the electrode to the base metal. What can be
   done to reduce this tendency while welding in the horizontal,
   vertical, and overhead positions?
   A. Increase the electrode diameter.
   B. Increase the welding current.
   C. Increase the arc gap slightly.
   D. Decrease the arc gap slightly.
   E. Increase the welding speed.

3. List two ways that can be successfully used to eliminate undercutting.
   _____
   _____
   _____

4. When a lap joint is welded, which piece of metal requires more heat?
   _____
   _____

Copyright by The Goodheart-Willcox Co., Inc.   Modern Welding Lab Workbook   79

5. A(n) _____° work angle is used to weld a horizontal groove weld.

5. _____

6. When a fillet weld is being made in a T-joint or inside corner joint in the horizontal position, the electrode should have a _____.
   A. 20° drag travel angle
   B. 20° push travel angle
   C. 30° push travel angle
   D. 45° drag travel angle

6. _____

7. The work angle for welding a horizontal fillet weld on a T-joint or inside corner joint is _____.

7. _____

8. A backward slanting weave motion is used for making a horizontal fillet weld in order to _____.
   A. add more metal to the trailing part of the weld pool
   B. create a shelf
   C. ensure proper penetration of both pieces
   D. allow the welder to see the trailing edge of the weld pool

8. _____

9. *True or False?* In SMAW, the travel angle used for a weld made in the horizontal position is the same as the travel angle used for a weld made in the flat position.

9. _____

10. A long arc length can cause _____ in a weld made in the horizontal position.
    A. lack of penetration
    B. slag inclusion
    C. overlap
    D. excessive penetration

10. _____

Job 6D-1

Fillet Weld on a Lap Joint in the Horizontal Welding Position

Name _____ Date _____

Class _____ Instructor _____

Learning Objective

● In this job, you will demonstrate your ability to weld a fillet weld on a lap joint in the horizontal welding position.

1. Obtain three pieces of mild steel that measure 1/4" × 1 1/2" × 5" (6.4mm × 40mm × 125mm) and six E6012 electrodes. (See Step 2A for the diameter.)
2. Answer the following questions before starting to weld. (Refer to Figure 6-17 in the text.)

   A. What is the smallest diameter electrode that is recommended for welding your metal? 2. A. _____

   B. What amperage range should be used? B. _____

3. Using AC and the electrode diameter and amperage settings previously described, tack weld the weldment in the flat position.
4. After tack welding, place the weldment into a weld positioning fixture or prop it against firebricks and make the welds shown in the following drawing.

   ■ Note: The weldment should be turned as necessary so all welds are made in the horizontal welding position.

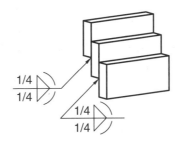

5. Read each weld prior to making the next. Make whatever changes are necessary to improve the next weld made.

Inspection

All fillet welds should be even in width, with evenly spaced ripples, and with no undercutting or overlapping.

Instructor's initials: _____

Job 6D-2

Fillet Weld on a T-Joint in the Horizontal Welding Position

Name _____ Date _____

Class _____ Instructor _____

> **Learning Objective**
> - In this job, you will weld fillet welds on a T-joint in the horizontal position.

1. Obtain three pieces of mild steel that measure 1/4" × 1 1/2" × 5" (6.4mm × 40mm × 125mm) and six E6011 electrodes. (See Step 2A for the diameter.)
2. Answer the following questions before beginning to weld. (Refer to Figure 6-17 in the text.)

    A. What is the largest diameter electrode recommended for this metal thickness?   2. A. _____

    B. What range of amperage should be used?   B. _____

    C. AC is not to be used! What DC polarity will be used? (See Figure 6-15 in the text.)   C. _____

3. Use the DC polarity, electrode diameter, and the amperage at the midpoint of the suggested amperage range shown in Step 2. Assemble and tack weld the weldment shown in the flat welding position.

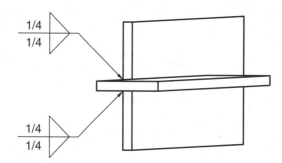

4. After tack welding, place the weldment in a weld positioning fixture or prop it against firebricks so that two joints are in the horizontal welding position. Weld two joints. Position the pieces so the remaining two joints are in the horizontal position. Complete the final two joints.
5. Read each weld prior to making the next. Make whatever changes are necessary to improve the next weld made.

> **Inspection**
> All fillet welds should have evenly spaced ripples, a uniform width, and have no undercutting or overlapping.
>
> Instructor's initials: _____

Job 6D-3

Bevel-Groove Weld on a Butt Joint and a V-Groove Weld on an Outside Corner Joint in the Horizontal Welding Position

Name _____ Date _____
Class _____ Instructor _____

> **Learning Objective**
> - In this job, you will be required to weld a bevel-groove butt joint and a V-groove outside corner joint in the horizontal welding position.

1. Obtain five pieces of mild steel that measure 1/4″ × 1 1/2″ × 5″ (6.4mm × 40mm × 125mm) and six 1/8″ (3.2mm) E6012 electrodes.
2. Answer the following questions before beginning to weld:

   A. What amperage range is recommended?   2. A. _____

   B. AC is not to be used! What polarity DC current is   B. _____
   suggested?

3. Bevel the correct edges to be used in the bevel groove butt joint shown in Step 6.
4. Tack weld all the pieces to create the weldment shown in Step 6. These tack welds should be made in the flat welding position.
5. After tack welding, place the weldment into a weld positioning fixture or lean it against firebricks.

   ■ **Note:** The weldment may be turned as required to make all welds in the horizontal welding position.

6. Complete all welds as required in the following welding drawing:

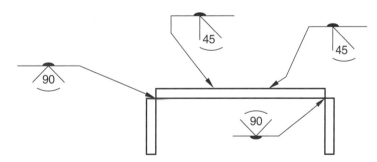

7. Read each weld before making the next one. Make whatever changes are necessary to improve the next weld made.

> **Inspection**
> All welds must have 100% penetration. The weld beads must have an even width, uniformly spaced ripples, be built-up, and be free of undercutting and overlapping.
>
>   Instructor's initials: _____

Copyright by The Goodheart-Willcox Co., Inc.   Modern Welding Lab Workbook   83

Lesson 6E

SMAW in the Vertical and Overhead Positions

Name _____ Date _____

Class _____ Instructor _____

Learning Objective
- You will be able to perform SMAW in the vertical and overhead positions.

Instructions
Carefully read Headings 6.9.6 through 6.9.7 of the text. Also study Figures 6-53 through 6-60 in the text. Then answer the following questions.

1. A(n) _____ direction of vertical welding is preferred to reduce slag inclusions and produce the strongest welds.

    1. _____

2. Why is a whipping motion used when performing welds in the horizontal, vertical, or overhead position?

    _____

    _____

3. *True or False?* When a vertical fillet weld on a lap joint is performed, the forward motion and any whipping motion is made on the surface, not on the part with the exposed edge.

    3. _____

4. Welds made in the _____ position are generally considered to be the most dangerous and the hardest to perform.

    4. _____

5. _____ must be worn for safety when welding in the overhead welding position.
    A. A hat or cap and an arc helmet
    B. Coveralls or work clothes without open pockets, buttoned at the collar
    C. Leather quality gloves
    D. A leather cape or coat
    E. All of the above.

    5. _____

6. For safety, the welder in the following figure is using a welding fume _____ and a powered _____ while welding a large assembly.

6. _____
   _____

7. Describe the steps and movements required to restart an arc and continue a bead.

   A. _____
   B. _____
   C. _____
   D. _____

8. *True or False?* Vertical welds should have the same appearance as a weld made in the flat welding position.

8. _____

9. During lap welding, when are the electrode and weld pool moved away from the edge to the surface of the metal?
   _____
   _____

10. List five ways to prevent filler metal sagging.
    _____
    _____
    _____
    _____
    _____
    _____
    _____

Job 6E-1

Fillet Weld on a Lap Joint in the Vertical Welding Position

Name _____ Date _____

Class _____ Instructor _____

Learning Objective

- In this job, you will use SMAW to make a fillet weld on a vertical lap joint.

▪ **Note:** Weak vertical welds are produced when the molten slag is included (trapped) in the weld.

1. Obtain three pieces of 3/32" (2.4mm) mild steel that measure 1 1/2" × 5" (40mm × 125mm) and six E6012 electrodes.
2. Answer the following questions before beginning to weld.

    A. What diameter electrodes should be used?　　2. A. _____

    B. What amperage range is suggested?　　　　　　B. _____

    C. What type current and polarity may be used?　　C. _____

    ▪ **Note:** See Figures 6-15 and 6-17 in the text.

3. Tack weld each joint to form the weldment shown in Step 5. The tack welds may be made in the flat welding position.
4. After tacking, place the weldment in a weld positioning fixture or prop it against firebricks.
5. Each fillet weld shown in the following drawing will be made in the vertical position.

    ▪ **Note:** Use the current indicated on the welding symbol.

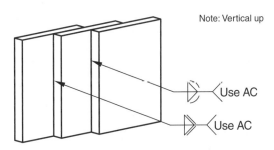

6. Read each weld prior to making the next weld. Make whatever changes are necessary to improve the next weld made.

Inspection

Each fillet weld must be convex or near flat in shape, have an even width, and evenly spaced ripples. The weld should have no undercutting or overlapping.

Instructor's initials: _____

Job 6E-2

Fillet Weld on a T-Joint in the Vertical Welding Position

Name _____ Date _____

Class _____ Instructor _____

Learning Objective

● In this job, you will demonstrate your ability to perform fillet welds on T-joints in the vertical welding position.

1. Obtain four pieces of mild steel that measure 1/4″ × 1 1/2″ × 5″ (6.4mm × 40mm × 125mm) and six E6013 electrodes.

2. Answer the following questions before beginning to weld:
   A. What is the smallest diameter electrode suggested?
   B. What is the current range suggested in Figure 6-17 in the text?
   C. What type of current and polarity is suggested?

   2. A. _____
      B. _____
      C. _____

3. Tack weld two weldments as shown in Step 5. The tack welding may be done in the flat position.

4. Place the weldment into a weld positioning fixture or lean it against firebricks so the joint is in the vertical position.

5. Make the welds shown in the following drawing in the vertical position.

   ■ Note: Make two of these weldments. Use the current and polarity shown on the weld symbol.

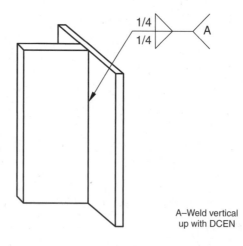

A—Weld vertical up with DCEN

Inspection

Each fillet weld should have an even width, smooth and even ripples, and be convex in shape. There should be no undercutting or overlapping.

Instructor's initials: _____

88 Modern Welding Lab Workbook Copyright by The Goodheart-Willcox Co., Inc.

Job 6E-3

V-Groove Weld on a Butt Joint in the Vertical Welding Position

Name _____ Date _____

Class _____ Instructor _____

> **Learning Objective**
> - In this job, you will learn how to perform a V-groove weld on a butt joint in the vertical welding position.

1. Obtain four pieces of mild steel that measure 1/4" × 1 1/2" × 5" (6.4mm × 40mm × 125mm) and six E6010 electrodes.
2. Before you start to weld, answer the following questions:

    A. What is the smallest diameter electrode suggested in Figure 6-17 in the text? 2. A. _____

    B. What current type and polarity is suggested in Figure 6-15 in the text? B. _____

    C. What is the suggested amperage range? C. _____

3. Wearing grinding goggles, grind, flame cut, or machine the edges of the metal as required by the weld symbols in Step 6.
4. Tack weld the metal to form the weldment shown in Step 6. Your tack welds can be made in the flat position.
5. After tack welding, place the weldment into a weld positioning fixture or lean it against firebricks.
6. Each weld must be made in the vertical welding position and according to the following drawing.

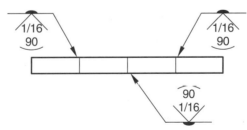

7. Read each weld before making the next weld. Make whatever changes are necessary to improve the next weld made.

> **Inspection**
> Each weld must have a 100% penetration showing on the root side. The weld beads must be even in width, have evenly spaced ripples, and have no undercutting or overlapping.
>
>   Instructor's initials: _____

Job 6E-4

Fillet Weld on a Lap Joint in the Overhead Welding Position

Name _____ Date _____

Class _____ Instructor _____

> **Learning Objective**
> - In this job, you will learn how to make a fillet weld in a lap joint in the overhead welding position.

1. Obtain three pieces of mild steel that measure 1/8″ × 1 1/2″ × 5″ (3.2mm × 40mm × 125mm) and four E6011 electrodes.
2. Refer to Figures 6-15 and 6-17 in the text to answer the following questions:

    A. What is the smallest diameter electrode recommended?   2. A. _____

    B. What type current and polarity is recommended?   B. _____

    C. What is the recommended current range?   C. _____

3. Tack each joint to form the weldment shown in Step 6. Your tack welding may be done in the flat welding position.
4. After tack welding, place the weldment into a weld positioning fixture at a height above your head.

    Caution: When welding in the overhead position, make certain to wear a cap, cape or leather jacket, gloves, and a helmet. Flash goggles should also be worn.

5. Make the welds shown in the following drawing in the overhead position. Use the type of current shown on the weld symbol. Turn the weldment over for the last two welds.

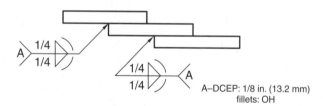

A–DCEP: 1/8 in. (13.2 mm)
fillets: OH

6. Read each weld before making the next weld. Make whatever changes are necessary to improve the next weld made.

> **Inspection**
> Your fillet welds must have a convex face, even ripples, and even width. There should be no overlapping or undercutting visible.

Instructor's initials: _____

Job 6E-5

Fillet Weld on a T-Joint in the Overhead Welding Position

Name _____ Date _____

Class _____ Instructor _____

> **Learning Objective**
> ● In this job, you will learn how to make a fillet weld on a T-joint in the overhead welding position.

1. Obtain three pieces of mild steel that measure 1/4" × 1 1/2" × 5" (6.4mm × 40mm × 125mm) and six E6012 electrodes.
2. Refer to Figures 6-15 and 6-17 in the text to answer the following questions:
   A. What is the smallest diameter electrode recommended? 2. A. _____
   B. What type current and polarity is recommended? B. _____
   C. What is the recommended current range? C. _____
3. Tack each joint to form the weldment shown in Step 6. The tack welding may be done in the flat position.
4. After tack welding, place the weldment into a weld positioning fixture and at a height above your head.

   ■ **Caution:** When welding in the overhead welding position, wear a cap, cape or leather jacket. Wear leather spats to cover your shoe tops. Also wear coveralls or work clothes buttoned at the collar, gloves, and a helmet. Flash goggles should also be worn.

5. Each fillet weld must be made in the overhead position. Turn the weldment over for the last two welds.
6. Make the welds shown in the following drawing in the overhead position and using AC.

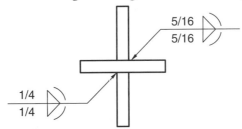

   ■ **Note:** It may be necessary to alter your travel speed or current settings to obtain a different size fillet on each side of the weldment.

7. Read each weld before making the next weld. Make whatever changes are necessary to improve the next weld made.

> **Inspection**
> Your fillet welds must have a convex face, even ripples, and even width. There should be no overlapping or undercutting visible.
>
> Instructor's initials: _____

Job 6E-6

Bevel-Groove Weld on a Butt Joint in the Overhead Welding Position

Name _____ Date _____

Class _____ Instructor _____

Learning Objective
- In this job, you will learn to make a bevel-groove weld in a butt joint in the overhead welding position.

1. Obtain three pieces of mild steel that measure 1/4" × 1 1/2" × 5" (6.4mm × 40mm × 125mm) and six E6010 electrodes.
2. Refer to Figures 6-15 and 6-17 in the text and answer the following questions:

    A. What is the smallest diameter electrode recommended?   2. A. _____

    B. What type current and polarity is recommended?   B. _____

    C. What is the recommended current range?   C. _____
3. Prepare the edges as necessary to create the joints shown in Step 6.
4. Tack each joint to form the weldment shown in Step 6. Your tack welding may be done in the flat position.
5. After tack welding, place the weldment into a weld positioning fixture and at a height above your head.
6. Each bevel-groove weld must be made in the overhead welding position.
7. Make the weldment shown in the following drawing. Perform all welds in the overhead position.

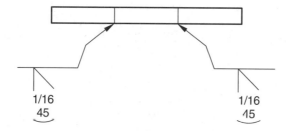

8. Read each weld before making the next weld. Make whatever changes are necessary to improve the next weld made.

Inspection
Your fillet welds must have a convex face, even ripples, and even width. There should be no overlapping or undercutting visible.

Instructor's initials: _____

92   Modern Welding Lab Workbook

Lesson 7A

GTAW Equipment and Supplies

Name _____ Date _____

Class _____ Instructor _____

Learning Objective
- You will be able to describe the functions of the welding machines used for GTAW. You will also be able to describe the shielding gases, accessory equipment, and electrodes used for GTAW.

Instructions
Carefully read the introduction to Chapter 7 and Headings 7.1 through 7.7 of the text. Study Figures 7-1 through 7-28 in the text. Then answer the following questions.

1. *True or False?* Arc welding machines used for GTAW can be either AC or DC, but must supply a constant current.

   1. _____

2. The curves shown in the following image are known as constant current curves, or _____ curves.

   2. _____

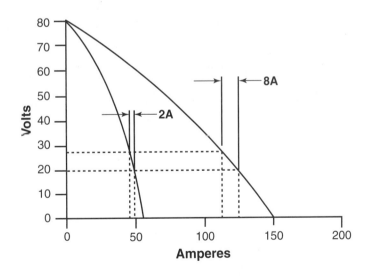

3. In gas tungsten arc welding, a high-frequency voltage is superimposed on the DC circuit for the purpose of _____ the arc.

   3. _____

4. In a constant current volt-ampere curve, if a large percentage change is made in the voltage, it will result in _____ percentage change in the current.
   A. a large
   B. no
   C. an extremely large
   D. a small
   E. less than 0.1

4. _____

5. High frequency is fed into the AC circuit during GTAW to stabilize the arc and to help start the arc. The high frequency _____.
   A. remains on only for starting
   B. remains on continuously
   C. is turned on and off as the AC changes
   D. remains on only when the amperage goes too high
   E. remains on only at low voltages

5. _____

6. During welding with AC, the arc tends to stop during the half cycle when the electrode becomes _____.
   A. negative
   B. positive
   C. neutral

6. _____

7. Name the three types of AC waves plotted in the following image.

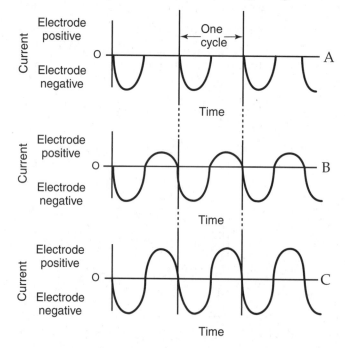

A. _____
B. _____
C. _____

Name _____

8. What causes one half of the AC curve to become rectified? 8. _____
   A. The amperage is too low to ionize the arc.
   B. The voltage cannot travel easily from the base metal surface to the small electrode surface.
   C. The current cannot easily travel from the base metal surface to the small electrode surface.
   D. The voltage is too high.
   E. The voltage cannot jump from point to point.

Use the following statements to answer Questions 9 and 10.

   A. Better oxide cleaning is provided because, during the electrode positive half of the cycle, more current flows.
   B. Can be generated with a less expensive welding machine.
   C. The arc is reliably reignited during the electrode positive half of the cycle.
   D. Capable of producing greater penetration.
   E. Excellent for high-quality production welding.
   F. Higher currents can be used with a given electrode.
   G. Produces a more stable arc.

9. List the letters of *all* of the statements that apply to a balanced AC wave. 9. _____

10. List the letters of *all* the statements that apply to an unbalanced AC wave. 10. _____

11. Which shielding gas provides easier arc starting, better metal cleaning action, and requires less volume to properly shield a weld? 11. _____
    A. Argon
    B. Helium
    C. CO_2
    D. Air
    E. Oxygen

12. *True or False?* Argon produces greater arc stability than helium for welding with AC. 12. _____

13. Mixing too much hydrogen with argon will cause ____. 13. _____
    A. too high a welding speed
    B. low available heat
    C. too much shielding
    D. porosity
    E. uneven heating

14. Regulators attached to inert gas cylinders usually have the outlet pressure preset at _____ psig.

    14. _____

    A. 20
    B. 30
    C. 40
    D. 50
    E. 60

15. Identify the parts of the device shown in the following image.

    A. _____
    B. _____
    C. _____
    D. _____

    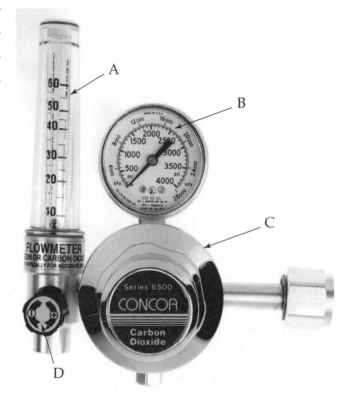

16. Name the parts of the gas-cooled, GTAW torch shown in the following image.

    A. _____
    B. _____
    C. _____
    D. _____
    E. _____

    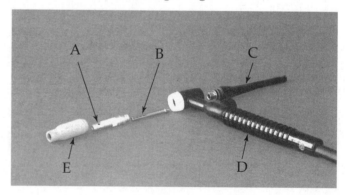

96   Modern Welding Lab Workbook

Lesson 7A GTAW Equipment and Supplies

Name _____

17. Within a GTAW torch, the electrode is held in place by the _____. Current travels to the electrode through the _____.

17. _____

18. On the lines provided, list the letters and numbers for each of the seven AWS classifications of tungsten electrodes, followed by the color painted on the end of each electrode.

   Pure tungsten _____

   1% thoriated tungsten _____

   2% thoriated tungsten _____

   Zirconiated tungsten _____

   Ceriated tungsten _____

   Lanthanum tungsten _____

   Other tungsten electrodes _____

19. *True or False?* Pliers can be used to break off the contaminated part of a tungsten electrode so the tip can be repointed.

19. _____

20. What is the greatest advantage of helium over argon?

_____

_____

Lesson 7B

GMAW and FCAW Equipment and Supplies

Name _____ Date _____
Class _____ Instructor _____

Learning Objective
- You will be able to set up a GMAW and FCAW station. You will also be able to select the electrode and shielding gas(es) and set the flow rate.

Instructions
Carefully read Headings 7.8 through 7.17 of the text. Also study Figures 7-29 through 7-66 in the text. Then answer the following questions.

1. GMAW is *not* a(n) _____ process. 1. _____
   A. manual
   B. semiautomatic
   C. mechanized
   D. automated

2. FCAW-S is an abbreviation for _____. 2. _____
   A. semiautomatic FCAW
   B. self-shielded FCAW
   C. solid wire FCAW
   D. silicon covered FCAW

3. Name the parts of the FCAW–S process shown in the following image.
   A. _____
   B. _____
   C. _____
   D. _____
   E. _____

Copyright by The Goodheart-Willcox Co., Inc. Modern Welding Lab Workbook 99

4. Which of the following is *not* a benefit of FCAW?
   A. Fluxing agents remove undesirable substances.
   B. Metallurgical qualities are controlled by adding alloying elements.
   C. A slag layer protects the hot weld bead.
   D. Steel plates over 1" (25mm) thick can be welded in a single pass.

4. _____

5. Name the parts of the GMAW station shown in the following image.

   A. _____
   B. _____
   C. _____
   D. _____
   E. _____
   F. _____
   G. _____

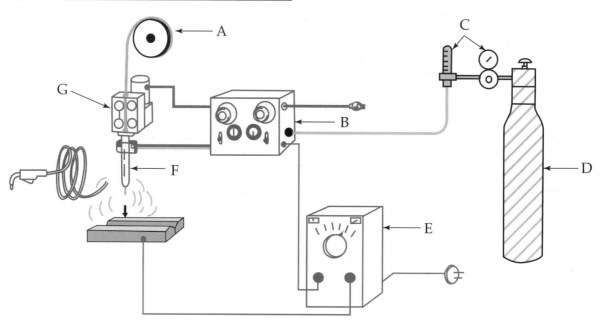

6. Power supplies used for GMAW and FCAW are constant ____ machines.

6. _____

7. Does a steeper slope on a machine's output increase or decrease the maximum current output of the machine?

7. _____

8. List four shielding gases used for GMAW.

_____

_____

9. List two shielding gases used for FCAW–G.

_____

Lesson 7B GMAW and FCAW Equipment and Supplies

Name _____

10. List two shielding gases that can be added to argon to reduce undercutting, improve flow, and provide a more stable arc in carbon and low alloy steels.
    _____

11. Which type of FCAW is more common, self-shielded or gas-shielded?
    _____

12. Which of the following is *not* turned on by pulling the trigger on a GMAW gun?   12. _____
    A. Welding current
    B. Shielding gas
    C. Power supply
    D. Wire feeder

13. Welding continuously or welding with argon or helium above _____ requires a water-cooled gun.   13. _____
    A. 150A
    B. 300A
    C. 450A
    D. 600A

14. List two reasons to use a pull gun.
    _____
    _____

15. Which of the following does *not* need to be replaced on occasion due to wear?   15. _____
    A. Contact tube
    B. Electrode lead
    C. Nozzle
    D. Liner

16. When is a nylon liner used?
    _____

17. Answer the following questions about an ER80S-6 electrode.
    What is the tensile strength in ksi?
    _____

    Is this a solid or tubular electrode?
    _____

    What does the -6 designate?
    _____

Modern Welding Lab Workbook

18. Answer the following questions about an E70T-4 electrode.
    What type shielding is to be used?
    _____

    Is this electrode used for single pass or multiple-pass welding?
    _____

    What polarity is to be used?
    _____

19. Answer the following questions about an E81T1-Ni2 electrode.
    Is this a solid or tubular electrode?
    _____

    What position(s) can this electrode be used in?
    _____

    Is this electrode used for single pass or multiple-pass welding?
    _____

    What polarity is to be used?
    _____

20. What lens shade number is recommended for GMAW welding with 200A?
    _____

Lesson 8A

Gas Tungsten Arc Welding Safety

Name _____ Date _____
Class _____ Instructor _____

Learning Objective
- You will be able to practice the safety procedures and precautions required when performing gas tungsten arc welding (GTAW).

Instructions
Carefully read Headings 7.1 through 7.7, 8.3, 8.3.5, 8.4, and 8.16 of the text. Then answer the following questions.

1. With the exception of hydrogen (H_2), the shielding gases used in GTAW are not combustible. What makes handling these cylinders of gas dangerous? _____
   _____
   _____

2. Which of the following statements does *not* apply to the safe handling of welding shielding gas cylinders? 2. _____
   A. The safety cap must be screwed on securely whenever cylinders are moved or stored.
   B. When in use or stored, cylinders must be securely fastened to a stable object.
   C. Care must be taken to avoid accidentally damaging cylinders with an arc or cutting torch.
   D. Cylinders should always be used and stored in the upright (vertical) position.
   E. Cylinders must be changed when they drop below 250 psig or 1700 kPa.

3. *True or False?* Oxygen and shielding gas cylinders should not be used with the cylinder valve partly open. 3. _____

4. *True or False?* If a label on a cylinder is not readable or is missing, do not assume that a cylinder contains a particular gas. The cylinder should be returned to the supplier. 4. _____

5. Shielding gases are used with a pressure regulator. The regulator screw should be ____ before the cylinder valve(s) are opened.
   A. screwed all the way in until the threads feel tight
   B. screwed out until the threads feel loose
   C. just 1/4 turn open
   D. half-open
   E. The position does not matter.

5. _____

6. Do not stand in front of the ____ as the cylinder valve is opened, even if it is opened slowly.

6. _____

7. For gas tungsten arc welding on ferrous metals with up to a 5/32″ (4mm) diameter electrode, which filter lens should be used?
   A. #6
   B. #8
   C. #10
   D. #11
   E. #12

7. _____

8. Unless you are wearing ____ breathing equipment, death from suffocation can occur if you enter an area that is filled with a shielding gas.

8. _____

9. All arc welding should be done in a booth or in an area shielded by curtains to protect others from arc ____.

9. _____

10. *True or False?* A welder should always wear safety goggles when cleaning metals.

10. _____

Lesson 8B

Gas Tungsten Arc Welding Principles

Name _____ Date _____
Class _____ Instructor _____

Learning Objective
You will be able to describe the principles of GTAW. You will also be able to select and set up the power source, and select the proper shielding gas, torch nozzle, tungsten electrode, and filler metal.

Instructions
Carefully read the introduction to Chapter 8 and Headings 8.1 through 8.3.7 of the text. Also study Figures 8-1 through 8-36 in the text. Then answer the following questions.

1. The GTAW process can be used to weld almost any metal. List the metals *not* popularly welded with the GTAW process. _____
   _____
   _____
   _____

2. *True or False?* Approximately two-thirds of the heat is generated on the workpiece when DCEP is used. 2. _____

3. Which DC polarity will provide the deepest penetration? 3. _____

4. ____ current is used when surface oxides must be removed. 4. _____
   A. AC or DCEP
   B. AC or DCEN
   C. Only DCEP
   D. Only DCEN
   E. Only AC

Copyright by The Goodheart-Willcox Co., Inc. Modern Welding Lab Workbook **105**

5. What type of current is shown in the following figure? 5. _____

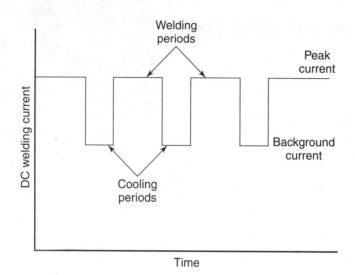

6. GTAW power sources that produce a(n) _____-wave AC output do not need high frequency to maintain the arc. 6. _____

7. A postflow timer on a GTAW power source _____. 7. _____
   A. shows open circuit voltage when no welding is occurring.
   B. sets the length of time a set current flows once the arc is initiated
   C. controls the duration of the pulse cycle
   D. controls how long the shielding gas flows after the arc is stopped

8. The following is a pulsed GTAW program. What is the purpose of the background current?

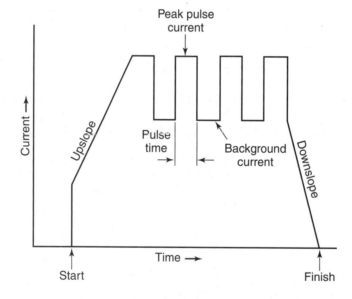

_____
_____
_____

Lesson 8B Gas Tungsten Arc Welding Principles

Name _____

9. Match the shielding gas on the left with the statement that describes it on the right.
   Helium
   Argon

   A. Cannot be used to produce acceptable welds when AC is used.
   B. Is better for use on thicker base metals because of its higher available heat.
   C. Requires a lower arc voltage and is preferred when welding thin metals.
   D. The addition of hydrogen to this shielding gas permits increased welding speeds for welding of stainless steel, nickel-copper, or nickel-based alloys.
   E. Does *not* produce a cleaning action on the base metal.

   9. A. _____
      B. _____
      C. _____
      D. _____
      E. _____

10. Argon and helium gas mixtures _____.
    A. produce a weld with deeper penetration than argon alone
    B. contain up to 25% helium
    C. provide better cleaning action than argon alone
    D. do not provide good cleaning action.

    10. _____

11. The suggested gas flow rate for any type of joint on 1/4″ (6.4mm) magnesium is _____ cfh (_____ L/min).

    11. _____
        _____

12. The exit diameter of a #10 nozzle is _____ inches (_____mm).

    12. _____
        _____

13. *True or False?* The choice of nozzle diameters is usually a compromise between adequate coverage of the entire weld area and access to the joint.

    13. _____

14. Tungsten electrodes are ground to a point or to a near point when used with _____ type current. _____ type electrodes are used with this type of current.

    14. _____
        _____

15. Which of the following electrodes is best suited for welding with DC?
    A. zirconiated
    B. pure
    C. thoriated
    D. Any type of tungsten electrode can be used for welding with DC.

    15. _____

16. Tungsten electrodes are ground to a point or to a near point for use with _____ type current.

    16. _____

17. Electrodes should be ground in a(n) _____ direction.

    17. _____

18. As a general rule, the electrode tip should ____ the end of the nozzle.
    A. be even with
    B. extend 1/2″ (12.5mm) beyond
    C. extend one electrode diameter beyond
    D. extend 1/4″ (6mm) beyond

18. _____

19. The color code for a zirconiated electrode is ____.
    A. black
    B. gray
    C. orange
    D. brown

19. _____

20. Define the term *downslope*. Why is downslope used?

    _____
    _____
    _____
    _____

Lesson 8C

Gas Tungsten Arc Welding Procedures

Name _____   Date _____

Class _____   Instructor _____

Learning Objective
● You will be able to prepare a weld. You will also be able to perform various welds in all positions using the GTAW process.

Instructions
Carefully read Headings 8.4 through 8.16 of the text. Also study Figures 8-37 through 8-68 in the text. Then answer the following questions.

1. Why are copper-coated steel welding rods not recommended for use with GTAW?
   _____
   _____

2. Identify the items shown in the following diagram of a high-frequency arc start.

   A. _____
   B. _____
   C. _____
   D. _____
   E. _____

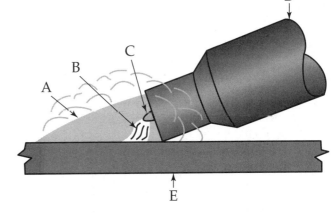

Copyright by The Goodheart-Willcox Co., Inc.   Modern Welding Lab Workbook   **109**

3. Identify and specify a range of values for the angles shown in the following drawing of GTAW in the flat position.

   A. _____

   B. _____

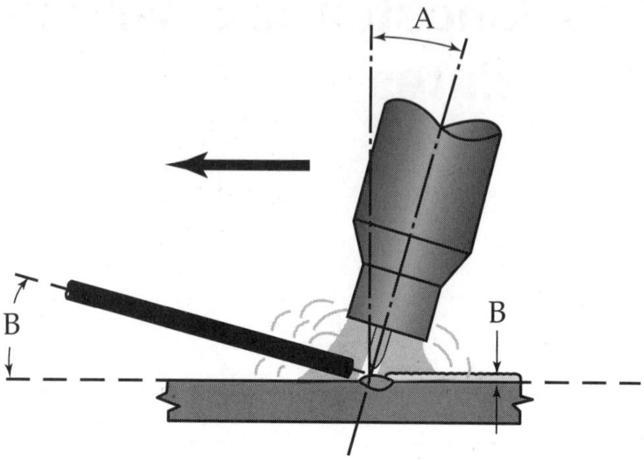

4. Why does the shielding gas automatically continue to flow for a short time after the GTAW arc is broken (stopped)?

   _____

   _____

   _____

   _____

5. GTAW is not normally used on carbon steel thicker than _____.   5. _____
   A. 1/8″ (3mm)
   B. 1/4″ (6mm)
   C. 3/8″ (9.5mm)
   D. 1/2″ (12.5mm)
   E. 5/8″ (16mm)

6. What can happen if the flowmeter is not closed completely when the GTAW station is shut down?

   _____

   _____

   _____

7. When the welding rod is added to the front of the weld pool, the torch is _____.   7. _____
   A. not moved until after the rod is removed
   B. moved toward the front of the crater
   C. moved toward the back of the crater
   D. moved to one side of the crater
   E. All of the above.

Lesson 8C   Gas Tungsten Arc Welding Procedures

Name _____

8. Specify a range of values for the angles shown in the following drawing of a lap joint being welded in the horizontal position.

   A. _____

   B. _____

   C. _____

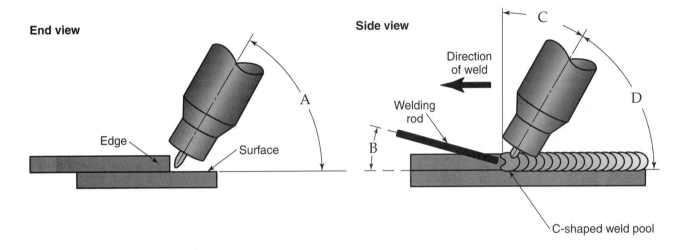

9. List three possible causes of rapid electrode consumption.

   _____

   _____

   _____

10. *True or False?* In a lap joint, the electrode should be pointed more towards the edge.

    10. _____

Copyright by The Goodheart-Willcox Co., Inc.   Modern Welding Lab Workbook   **111**

Job 8C-1

Fillet Weld on a Lap Joint in the Flat Welding Position

Name _____ Date _____

Class _____ Instructor _____

Learning Objective
● In this job, you will perform a fillet weld on a lap joint in the flat welding position.

1. Obtain three pieces of 1/16" (1.6mm) mild steel that measure 1 1/2" × 5" (40mm × 125mm).

2. Thoroughly clean oil, rust, and dirt from these pieces.

3. The following questions must be answered before you start to weld. Refer to Figures 8-19 and 8-20 in the text.

   A. What diameter electrode should be used?   3. A. _____
   B. What diameter welding rod will be used?   B. _____
   C. What amperage range will be used?   C. _____
   D. What type of shielding gas is to be used?   D. _____
   E. The rate of gas flow is _____ cfh (_____ L/min).   E. _____
   F. What type of current is to be used?   F. _____
   G. What type of tungsten electrode is to be used?   G. _____
   H. What electrode tip shape should be used?   H. _____

4. Use a #6–#8, 3/8"–1/2" (9.5mm–12.7mm) diameter nozzle for this job. Install the torch nozzle and prepare the electrode. Place the electrode into the torch using the proper size collet and the proper extension for the electrode.

5. Set the machine for the proper current and polarity.

6. Using the correct shielding gas, start the gas flow. Set the correct flow rate on the flowmeter.

7. Tack weld the metal pieces to form the weldment shown in the following figure.

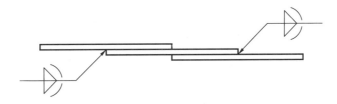

8. Place the weldment into a weld positioning fixture or lean it against firebricks. Weld all the fillets in the flat welding position. Turn the metal as required to weld in the flat welding position.

   ■ **Note:** "Read" each weld and change whatever is required to improve the quality of the next weld.

Inspection

Your fillet welds should be convex and even in width. The welds should have evenly spaced ripples, with no overlap or undercut.

Instructor's initials: _____

Job 8C-2

Fillet Weld on an Inside Corner in the Flat Welding Position

Name _____ Date _____

Class _____ Instructor _____

> **Learning Objective**
> ● In this job, you will demonstrate your ability to weld a fillet weld on an inside corner joint in the flat welding position.

1. Obtain the following mild steel pieces.
   A. Two pieces 1/8" × 1 1/2" × 5" (3.2mm × 40mm × 125mm).
   B. One piece 1/8" × 3" × 5" (3.2mm × 75mm × 125mm).

2. Thoroughly clean all pieces.

3. The following questions must be answered before you start to weld. Refer to Figures 8-19 and 8-20 in the text.
   A. What diameter electrode should be used?   3. A. _____
   B. What diameter welding rod will be used?   B. _____
   C. What amperage range will be used?   C. _____
   D. What type of shielding gas is to be used?   D. _____
   E. The rate of gas flow is ____ cfh (____ L/min).   E. _____
   F. What type of current is to be used?   F. _____
   G. What type of tungsten electrode is to be used?   G. _____
   H. What electrode tip shape should be used?   H. _____

4. Use a #6–#8 torch nozzle for this job. Install the torch nozzle and prepare the electrode. Place the electrode into the torch using the proper collet diameter and electrode extension.

5. Set the machine for the proper current and polarity.

6. Using the correct shielding gas, start the gas flow. Set the flow rate on the flowmeter.

7. Tack weld the pieces to form the weldment shown in the following figure.

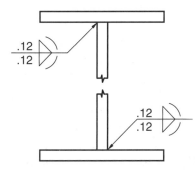

8. After tack welding, place the weldment into a weld positioning fixture or lean it against firebricks.

9. Weld all fillets in the flat welding position. Turn the metal as required to weld in the flat welding position.

   ▎**Note:** "Read" each weld and change whatever is required to improve the quality of the next weld.

Inspection
Your fillet welds should be convex and even in width. The welds should have evenly spaced ripples, with no overlap or undercut.

Instructor's initials: _____

Job 8C-3

Square-Groove Weld on a Butt Joint in the Flat Welding Position

Name _____ Date _____

Class _____ Instructor _____

Learning Objective

In this job, you will weld a square-groove weld on a butt joint in the flat welding position.

1. Obtain five pieces of mild steel that measure 1/8″ × 1 1/2″ × 5″ (3.2mm × 40mm × 125mm).

2. Thoroughly clean the pieces.

3. The following questions must be answered before you start to weld. Refer to Figures 8-19 and 8-20 in the text.

    A. What diameter electrode should be used?
    B. What diameter welding rod will be used?
    C. What amperage range will be used?
    D. What type of shielding gas is to be used?
    E. The rate of gas flow is _____ cfh (_____ L/min).
    F. What type of current is to be used?
    G. What type of tungsten electrode is to be used?
    H. What electrode tip shape should be used?

    3. A. _____
    B. _____
    C. _____
    D. _____
    E. _____
    F. _____
    G. _____
    H. _____

4. Use a #6–#8 diameter nozzle for this job. Install the torch nozzle and prepare the electrode.

5. Set the machine for the proper current and polarity.

6. Using the correct shielding gas, start the gas flow. Set the correct flow rate on the flowmeter.

7. Tack weld the metal pieces to form the weldment shown in the following figure.

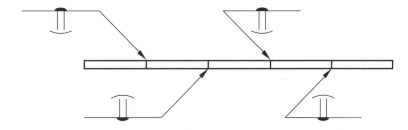

Copyright by The Goodheart-Willcox Co., Inc. Modern Welding Lab Workbook 117

8. Place the weldment into a weld positioning fixture or place it on the welding table.

9. Weld all the square-groove butt welds shown in the flat welding position.

   ■ **Note:** "Read" each weld and change whatever is required to improve the quality of the next weld.

Inspection

Your welds should have a convex face, with even width and evenly spaced ripples in the bead. All welds should show 100% penetration on the root side of the joint.

Instructor's initials: _____

Job 8C-4

Fillet Weld on a Lap Joint in the Horizontal Welding Position

Name _____ Date _____

Class _____ Instructor _____

Learning Objective
In this job, you will produce fillet welds on a lap joint in the horizontal welding position.

1. Obtain three pieces of mild steel that measure 1/8″ × 1 1/2″ × 5″ (3.2mm × 40mm × 125mm).

2. Thoroughly clean oil, rust, and dirt from these pieces.

3. The following questions must be answered before you start to weld. Refer to Figures 8-19 and 8-20 in the text.

   A. What diameter electrode should be used?   3. A. _____
   B. What diameter welding rod will be used?   B. _____
   C. What amperage range will be used?   C. _____
   D. What type of shielding gas is to be used?   D. _____
   E. The rate of gas flow is _____ cfh (_____ L/min). E. _____
   F. What type of current is to be used?   F. _____
   G. What type of tungsten electrode is to be used? G. _____
   H. What electrode tip shape should be used?   H. _____

4. Use a #6–#8 diameter nozzle for this job. Install the torch nozzle and prepare the electrode. Place the electrode into the torch using the proper size collet and the proper extension for the electrode.

5. Set the machine for the proper current and polarity.

6. Using the correct shielding gas, start the gas flow. Set the correct flow rate on the flowmeter.

7. Tack weld the metal pieces to form the weldment shown in the following figure.

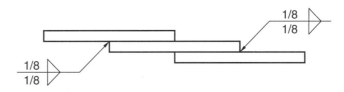

8. Place your weldment into a weld positioning fixture or support it with firebricks.

9. Weld all the fillets shown in the horizontal welding position. Adjust the weldment's position as needed so that each weld can be made in the horizontal welding position.

   ■ **Note:** "Read" each weld and change whatever is required to improve the quality of the next weld.

Inspection

Your fillet welds should be convex and even in width. The welds should have evenly spaced ripples, with no overlap or undercut.

Instructor's initials: _____

Job 8C-5

Fillet Weld on an Inside Corner and a T-Joint in the Horizontal Welding Position

Name _____ Date _____

Class _____ Instructor _____

> **Learning Objective**
> - In this job, you will make a fillet weld on an inside corner and T-joint in the horizontal welding position. A butt weld will also be made.

1. Obtain mild steel with the following sizes:
   Four pieces: 1/16" × 1 1/2" × 5" (1.6mm × 40mm × 125mm).
   One piece: 1/16" × 5" × 6" (1.6mm × 125mm × 150mm).

2. Thoroughly clean oil, rust, and dirt from these pieces.

3. The following questions must be answered before you start to weld. Refer to Figures 8-19 and 8-20 in the text.

   A. What diameter electrode should be used? 3. A. _____
   B. What diameter welding rod will be used?   B. _____
   C. What amperage range will be used?   C. _____
   D. What type of shielding gas is to be used?   D. _____
   E. The rate of gas flow is _____ cfh (_____ L/min). E. _____
   F. What type of current is to be used?   F. _____
   G. What type of tungsten electrode is to be used? G. _____
   H. What electrode tip shape should be used?   H. _____

4. Use a #6–#8 diameter nozzle for this job. Install the torch nozzle and prepare the electrode. Place the electrode into the torch using the proper size collet and the proper extension for the electrode.

5. Set the machine for the proper current and polarity.

6. Using the correct shielding gas, start the gas flow. Set the correct flow rate on the flowmeter.

7. Study the following figure. Decide which piece to tack in place and weld first. Which piece will be tacked and welded second, and then third? Which piece will be attached last? Think about torch access to each weld joint.

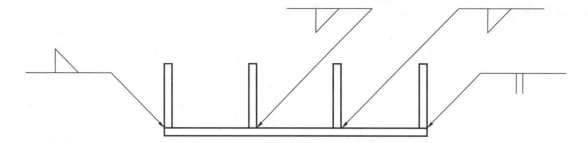

8. Tack weld the first piece in place. Weld the joint in the horizontal welding position. Add the filler metal to the upper edge of the weld pool.

9. Tack weld the second piece in place and weld in the horizontal position.

10. Repeat Step 9 with the third and fourth pieces.

   ▌ **Note:** "Read" each weld and change whatever is required to improve the quality of the next weld.

Inspection

Your fillet welds and butt weld should be convex and even in width. The welds should have evenly spaced ripples, with no overlap or undercut.

Instructor's initials: _____

Job 8C-6

Square-Groove Weld on a Butt Joint in the Horizontal Welding Position

Name _____ Date _____

Class _____ Instructor _____

Learning Objective
- In this job, you will demonstrate your ability to make a square-groove weld on a butt joint in the horizontal welding position.

1. Obtain five pieces of mild steel that measure 1/8" × 1 1/2" × 5" (3.2mm × 40mm × 125mm).

2. Thoroughly clean these pieces.

3. The following questions must be answered before you start to weld. Refer to Figures 8-19 and 8-20 in the text.

   A. What diameter electrode should be used? 3. A. _____
   B. What diameter welding rod will be used?   B. _____
   C. What amperage range will be used?   C. _____
   D. What type of shielding gas is to be used?   D. _____
   E. The rate of gas flow is _____ cfh (_____ L/min). E. _____
   F. What type of current is to be used?   F. _____
   G. What type of tungsten electrode is to be used? G. _____
   H. What electrode tip shape should be used?   H. _____

4. Use a #6–#8 diameter nozzle for this job. Install the torch nozzle and prepare the electrode. Place the electrode into the torch using the proper size collet and the proper extension for the electrode.

5. Set the machine for the proper current and polarity.

6. Using the correct shielding gas, start the gas flow. Set the correct flow rate on the flowmeter.

7. Set a 1/16" (1.6mm) root opening. Tack weld the metal pieces to form the weldment shown in the following figure.

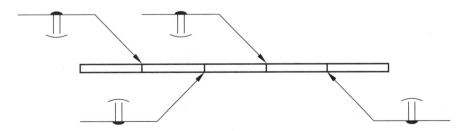

8. Place the weldment into a weld positioning fixture or lean it against firebricks.

9. Weld all square-groove butt welds in the horizontal welding position. Form and maintain a keyhole while welding. Add the filler metal to the upper edge of the weld pool. Point the electrode slightly upward.

   Note: "Read" each weld and change whatever is required to improve the quality of the next weld.

Inspection

The welds should have a convex face, with an even width, and evenly spaced ripples in the bead. All welds should show 100% penetration on the root side of the joint.

Instructor's initials: _____

Job 8C-7

Fillet Weld on a Lap Joint in the Vertical Welding Position

Name _____ Date _____

Class _____ Instructor _____

> **Learning Objective**
> - In this job, you will perform a fillet weld on a lap joint in the vertical welding position.

1. Obtain three pieces of mild steel that measure 1/8″ × 1 1/2″ × 5″ (3.2mm × 40mm × 125mm).

2. Thoroughly clean all pieces.

3. The following questions must be answered before you start to weld. Refer to Figures 8-19 and 8-20 in the text.
   A. What diameter electrode should be used?
   B. What diameter welding rod will be used?
   C. What amperage range will be used?
   D. What type of shielding gas is to be used?
   E. The rate of gas flow is _____ cfh (_____ L/min).
   F. What type of current is to be used?
   G. What type of tungsten electrode is to be used?
   H. What electrode tip shape should be used?

   3. A. _____
      B. _____
      C. _____
      D. _____
      E. _____
      F. _____
      G. _____
      H. _____

4. Use a #6–#8 diameter nozzle for this job. Install the torch nozzle and prepare the electrode. Place the electrode into the torch using the proper size collet and the proper extension for the electrode.

5. Set the machine for the proper current and polarity. Use a pulsed arc if the machine is equipped to furnish a pulsed arc.

6. Using the correct shielding gas, start the gas flow. Set the correct flow rate on the flowmeter.

7. Tack weld the pieces to form the weldment shown in the following figure.

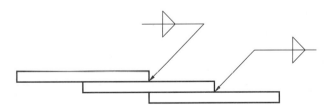

8. Place the weldment into a weld positioning fixture or lean it against firebricks.

   Should a vertical weld on metal greater than 1/16" (1.6mm) thick be made uphill or downhill?
   _____

9. Weld all fillet welds in the vertical welding position. Adjust the weldment's position as needed so that each weld can be made in the vertical welding position.

   ■ **Note:** "Read" each weld and change whatever is required to improve the quality of the next weld.

Inspection

Your fillet welds should be convex and even in width. The welds should have evenly spaced ripples, with no overlap or undercut.

Instructor's initials: _____

Job 8C-8

Fillet Weld on a T-Joint in the Vertical Welding Position

Name _____ Date _____

Class _____ Instructor _____

Learning Objective
- In this job, you will perform a fillet weld on a T-joint in the vertical welding position.

1. Obtain three pieces of mild steel that measure 1/16″ × 1 1/2″ × 5″ (1.6mm × 40mm × 125mm).

2. Thoroughly clean these pieces.

3. The following questions must be answered before you start to weld. Refer to Figures 8-19 and 8-20 in the text.
    A. What diameter electrode should be used? 3. A. _____
    B. What diameter welding rod will be used?   B. _____
    C. What amperage range will be used?   C. _____
    D. What type of shielding gas is to be used?   D. _____
    E. The rate of gas flow is _____ cfh (_____ L/min). E. _____
    F. What type of current is to be used?   F. _____
    G. What type of tungsten electrode is to be used? G. _____
    H. What electrode tip shape should be used?   H. _____

4. Use a #6–#8 diameter nozzle for this job. Install the torch nozzle and prepare the electrode. Place the electrode into the torch using the proper size collet and the proper extension for the electrode.

5. Set the machine for the proper current and polarity. Use a pulsed arc if the machine is equipped to furnish a pulsed arc.

6. Using the correct shielding gas, start the gas flow. Set the correct flow rate on the flowmeter.

7. Tack weld the metal pieces to form the weldment shown in the following figure.

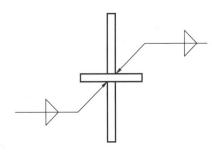

8. Place the weldment into a weld positioning fixture or lean it against firebricks.

9. Weld two of the weldments, making all welds in the vertical welding position.

   ■ **Note:** "Read" each weld and change whatever is required to improve the quality of the next weld.

Inspection

Your fillet welds should be convex and even in width. The welds should have evenly spaced ripples, with no overlap or undercut.

Instructor's initials: _____

Job 8C-9

Square-Groove Weld on a Butt Joint in the Vertical Welding Position

Name _____ Date _____

Class _____ Instructor _____

> **Learning Objective**
>
> • In this job, you will perform a square-groove weld on a butt joint in the vertical welding position.

1. Obtain five pieces of mild steel that measure 1/16″ × 1 1/2″ × 5″ (1.6mm × 40mm × 125mm).

2. Thoroughly clean these pieces.

3. The following questions must be answered before you start to weld. Refer to Figures 8-19 and 8-20 in the text.

   A. What diameter electrode should be used? 3. A. _____
   B. What diameter welding rod will be used? B. _____
   C. What amperage range will be used? C. _____
   D. What type of shielding gas is to be used? D. _____
   E. The rate of gas flow is _____ cfh (_____ L/min). E. _____
   F. What type of current is to be used? F. _____
   G. What type of tungsten electrode is to be used? G. _____
   H. What electrode tip shape should be used? H. _____

4. Use a #6–#8 diameter nozzle for this job. Install the torch nozzle and prepare the electrode. Place the electrode into the torch using the proper size collet and the proper extension for the electrode.

5. Set the machine for the proper current and polarity. Use a pulsed arc if the machine is equipped to furnish a pulsed arc.

6. Using the correct shielding gas, start the gas flow. Set the correct flow rate on the flowmeter.

7. Tack weld the metal pieces to form the weldment shown in the following figure.

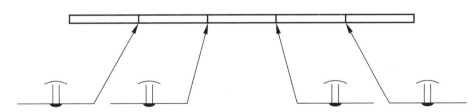

8. Place the weldment into a weld positioning fixture or lean it against firebricks so that all welds are performed in the vertical welding position.

9. Weld all welds in the vertical welding position.

   ▮ **Note:** "Read" each weld and change whatever is required to improve the quality of the next weld.

Inspection

Your welds should have a convex face, with an even width, and evenly spaced ripples in the bead. All welds should show 100% penetration on the root side of the joint.

Instructor's initials: _____

Job 8C-10

Fillet Weld on a Lap Joint in the Overhead Welding Position

Name _____ Date _____

Class _____ Instructor _____

Learning Objective
- In this job, you will perform a fillet weld on a lap joint in the overhead welding position.

1. Obtain three pieces of 1/16" (1.6mm) mild steel that measure 1 1/2" × 5" (40mm × 125mm).

2. Thoroughly clean oil, rust, and dirt from these pieces.

3. The following questions must be answered before you start to weld. Refer to Figures 8-19 and 8-20 in the text.

    A. What diameter electrode should be used?   3. A. _____
    B. What diameter welding rod will be used?   B. _____
    C. What amperage range will be used?   C. _____
    D. What type of shielding gas is to be used?   D. _____
    E. The rate of gas flow is ____ cfh (____ L/min). E. _____
    F. What type of current is to be used?   F. _____
    G. What type of tungsten electrode is to be used? G. _____
    H. What electrode tip shape should be used?   H. _____

4. Use a #6–#8 (3/8" or 9.5mm–12.7mm) diameter nozzle for this job. Install the torch nozzle and prepare the electrode. Place the electrode into the torch using the proper size collet and the proper extension for the electrode.

5. Set the machine for the proper current and polarity. Use a pulsed arc if the machine is equipped to furnish a pulsed arc.

6. Using the correct shielding gas, start the gas flow. Set the correct flow rate on the flowmeter.

7. Tack weld the metal pieces to form the weldment shown in the following figure.

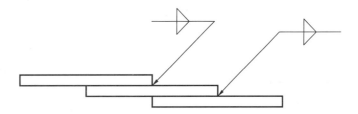

8. Place the weldment into a weld positioning fixture.

9. Weld all fillet welds in the overhead welding position. Turn the metal as required.

   ■ **Note:** "Read" each weld and change whatever is required to improve the quality of the next weld.

Inspection

Your fillet welds should be convex and even in width. The welds should have evenly spaced ripples, with no overlap or undercut.

Instructor's initials: _____

Job 8C-11

Fillet Weld on a T-Joint in the Overhead Welding Position

Name _____ Date _____

Class _____ Instructor _____

> **Learning Objective**
> - In this job, you will demonstrate your ability to perform a fillet weld on a T-joint in the overhead welding position.

1. Obtain three pieces of mild steel that measure 1/8″ × 1 1/2″ × 5″ (3.2mm × 40mm × 125mm).

2. Thoroughly clean oil, rust, and dirt from these pieces.

3. The following questions must be answered before you start to weld. Refer to Figures 8-19 and 8-20 in the text.

   A. What diameter electrode should be used?
   B. What diameter welding rod will be used?
   C. What amperage range will be used?
   D. What type of shielding gas is to be used?
   E. The rate of gas flow is _____ cfh (_____ L/min).
   F. What type of current is to be used?
   G. What type of tungsten electrode is to be used?
   H. What electrode tip shape should be used?

   3. A. _____
      B. _____
      C. _____
      D. _____
      E. _____
      F. _____
      G. _____
      H. _____

4. Use a #6–#8 diameter nozzle for this job. Install the torch nozzle and prepare the electrode. Place the electrode into the torch using the proper size collet and the proper extension for the electrode.

5. Set the machine for the proper current and polarity. Use a pulsed arc if the machine is equipped to furnish a pulsed arc.

6. Using the correct shielding gas, start the gas flow. Set the correct flow rate on the flowmeter.

7. Tack weld the metal pieces to form the weldment shown in the following figure.

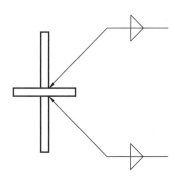

Modern Welding Lab Workbook 133

8. Place the weldment into a weld positioning fixture.

9. Weld all fillets in the overhead welding position. Turn the metal as required.

   ■ **Note:** "Read" each weld and change whatever is required to improve the quality of the next weld.

Inspection

Your fillet welds should be convex and even in width. The welds should have evenly spaced ripples, with no overlap or undercut.

Instructor's initials: _____

Job 8C-12

Square-Groove Weld on a Butt Joint in the Overhead Welding Position

Name _____ Date _____
Class _____ Instructor _____

> **Learning Objective**
> In this job, you will perform a square-groove weld on a butt joint in the overhead welding position.

1. Obtain five pieces of mild steel that measure 1/16″ × 1 1/2″ × 5″ (1.6mm × 40mm × 125mm).

2. Thoroughly clean oil, rust, and dirt from these pieces.

3. The following questions must be answered before you start to weld. Refer to Figures 8-19 and 8-20 in the text.

    A. What diameter electrode should be used?   3. A. _____
    B. What diameter welding rod will be used?   B. _____
    C. What amperage range will be used?   C. _____
    D. What type of shielding gas is to be used?   D. _____
    E. The rate of gas flow is _____ cfh (_____ L/min).   E. _____
    F. What type of current is to be used?   F. _____
    G. What type of tungsten electrode is to be used?   G. _____
    H. What electrode tip shape should be used?   H. _____

4. Use a #6–#8 diameter nozzle for this job. Install the torch nozzle and prepare the electrode. Place the electrode into the torch using the proper size collet and the proper extension for the electrode.

5. Set the machine for the proper current and polarity. Use a pulsed arc if the machine is equipped to furnish a pulsed arc.

6. Using the correct shielding gas, start the gas flow. Set the correct flow rate on the flowmeter.

7. Tack weld the metal pieces to form the weldment shown in the following figure.

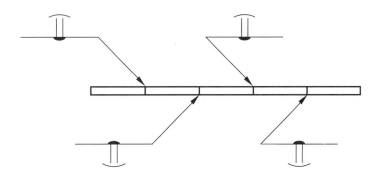

8. Place the weldment into a weld positioning fixture so that all welds are performed in the overhead welding position.

9. Weld all welds shown in the overhead welding position.

   ■ **Note:** "Read" each weld and change whatever is required to improve the quality of the next weld.

Inspection

Your welds should have a convex face, with an even width, and evenly spaced ripples in the bead. All welds should show 100% penetration on the root side of the joint.

Instructor's initials: _____

Job 8C-13

Groove Welds on Stainless Steel

Name _____ Date _____

Class _____ Instructor _____

> **Learning Objective**
>
> • In this job, you will perform a V-groove weld on a butt joint on stainless steel in the 1G (flat) and 2G (horizontal) welding positions.

1. Obtain eight pieces of stainless steel that measure 1/4" × 1 1/2" × 5" (6.4mm × 40mm × 125mm).

2. The following questions must be answered before you start to weld. Refer to Figures 8-19 and 8-20 in the text.

    A. What is the best polarity for use with GTAW on stainless steel?
    B. What diameter tungsten electrode is suggested?
    C. What diameter stainless welding wire is suggested?
    D. What amperage range will be used?
    E. What is the suggested shielding gas used on stainless steel?
    F. What gas flow rate should be set on the flowmeter?

    3. A. _____
       B. _____
       C. _____
       D. _____
       E. _____
       F. _____

3. Grind, arc cut, or machine the edges of the pieces to make the joints shown in the following image.

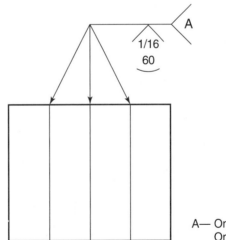

A— One in 1G position;
One in 2G position

Modern Welding Lab Workbook **137**

4. Use a #8–#10 diameter nozzle for this job. Install the torch nozzle and prepare the electrode. Place the electrode into the torch using the proper size collet and the proper extension for the electrode.

5. Tack weld four pieces as shown in the drawing in Step 3.

6. Place the weldment into a weld positioner and weld in the 1G position. More than one weld pass will be required to fill the joint. "Read" each weld and make any required changes before making the next weld.

7. Tack weld the remaining four pieces as shown in the drawing in Step 3.

8. Place these pieces in a weld positioner and weld in the 2G position. "Read" each weld and change whatever is required to improve the quality of the next weld.

Inspection

Each weld must have 100% penetration shown on the root side. The weld beads must be even in width, have smooth evenly spaced ripples, and no overlapping or undercutting.

Instructor's initials: _____

Lesson 9A

Gas Metal and Flux Cored Arc Welding Safety

Name _____ Date _____

Class _____ Instructor _____

Learning Objective
- You will be able to practice the safety procedures and precautions required when performing gas metal and flux cored arc welding.

Instructions
Carefully review Headings 7.14 through 7.17, Heading 9.16, and the safety warnings in Headings 9.4.6, 9.7, and 9.12 of the text. Then answer the following questions.

1. Why should you wear dark clothing while welding?
   _____
   _____

2. Removal of carbon monoxide and ozone from the weld area can be accomplished with a GMAW gun equipped with a(n) _____.
   2. _____

3. When a welding station is inspected before welding, all _____, _____, and _____ connections must be checked for safety.
   3. _____

4. You should not touch energized parts of the welding gun with bare skin or _____ gloves.
   4. _____

5. When compressed air is used to clean out an empty GMAW liner, where should the liner be pointed?
   _____
   _____

6. When CO_2 is used as a shielding gas for GMAW, _____ gas is generated.
   6. _____

7. *True or False?* Ozone is a highly toxic gas that is produced during gas metal arc welding.
   7. _____

Copyright by The Goodheart-Willcox Co., Inc.   Modern Welding Lab Workbook   139

8. What protective clothing should a welder wear for overhead welding?

_____

9. For long welding sessions, flash goggles with a #____ lens shade should be worn under an arc helmet.

9. _____

10. Clothing made from ____ provides the best protection from burns.

10. _____

Lesson 9B

Gas Metal and Flux Cored Arc Welding Principles

Name _____ Date _____

Class _____ Instructor _____

Learning Objective

You will be able to describe the principles of GMAW and FCAW.

Instructions

Carefully read the introduction to Chapter 9 and Headings 9.1 through 9.2.4 of the text. Also study Figures 9-1 through 9-9 in the text. Then answer the following questions.

1. *True or False?* GMAW deposits less weld metal in lbs/hr (kg/hr) than SMAW and GTAW.
   1. _____

2. *True or False?* GMAW is done with DCEP (DCRP) only, except for some surfacing applications.
   2. _____

3. The percentage of electrode metal deposited during welding with a flux cored electrode is ____% to ____%.
   3. _____

4. List four ways in which metal can be transferred in the GMAW process.

   _____

   _____

   _____

   _____

5. In GMAW–S, a(n) ____ force around the electrode squeezes the molten end of the electrode.
   5. _____

6. Which metal transfer method uses the lowest welding current? _____

   _____

7. Which GMAW metal transfer method uses the highest welding current? _____

   _____

8. What metal transfer method is shown in the following image?

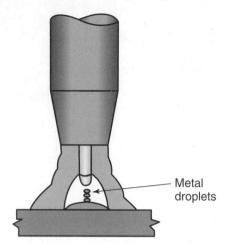

_____

9. What is/are the advantage(s) of using the short-circuiting method?
   A. It works well on thick metal sections.
   B. There is no metal transfer through the arc.
   C. It works well in all welding positions.
   D. It works well on thin metals.
   E. All of the above.

   9. _____

10. Short-circuiting of an electrode can cause the current to rise from 150A to 500A. What is used to slow down the possible rapid rise of the welding current? _____

_____

11. When the globular transfer method is used, _____ causes the metal droplet to leave the electrode wire.
    A. magnetism
    B. high currents
    C. high voltages
    D. short arc
    E. the weight of the droplet

    11. _____

12. The globular transfer method and a short arc length will cause a buried or _____ arc to occur.

    12. _____

13. The spray transfer method will deposit metal at _____ lbs/hr compared to _____ lbs/hr with the short-circuiting method.

    13. _____

Lesson 9B Gas Metal and Flux Cored Arc Welding Principles

Name _____

14. Assume a welder earns $10.00/hour and a weldment requires 60 lbs of metal.

   ■ **Note:** Use the highest lbs/hr figure given in the GMAW deposition rates table (Figure 9-3). Show your math in the space provided.

   A. What would be the total labor cost of the welder if he or she uses the short-circuiting method to complete the weldment?   14. A. _____

   B. What would be the total labor cost of the welder if he or she uses the spray transfer method to complete the weldment?   B. _____

15. *True or False?* All spray transfer methods occur above the transition current. The transition current varies with the electrode diameter, the electrode composition, and the electrode extension.   15. _____

16. The approximate transition current for a 0.045" diameter aluminum electrode is _____ amperes.   16. _____

17. The pulsed spray transfer method uses a peak current for welding. The background current is used to _____ the arc.   17. _____

18. List the letters of any of the following that are advantages of the pulsed spray transfer method.   18. _____
    A. Higher current levels make it possible to weld out of position.
    B. Thin metal sections can be welded.
    C. Very little spatter is created.
    D. Larger diameter electrodes can be used.

19. *True or False?* The pulsed spray transfer method can be used in auto repair shops to weld light steel parts with very low heat inputs.   19. _____

20. Most shielding gas mixtures used for spray transfer contain at least _____% argon.   20. _____

Lesson 9C

Gas Metal and Flux Cored Arc Welding—Setting Up the Station

Name _____ Date _____

Class _____ Instructor _____

Learning Objective
- You will be able to set up a GMAW/FCAW station. You will also be able to select the electrode and shielding gas(es) and set the flow rate.

Instructions
Carefully read Headings 9.3 through 9.4.7 of the text. Also study Figures 9-10 through 9-40 in the text. Then answer the following questions.

1. Identify the components labeled in the GMAW outfit diagram shown.

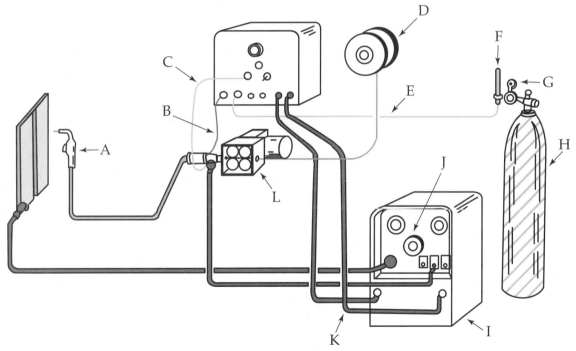

A. _____   E. _____   I. _____
B. _____   F. _____   J. _____
C. _____   G. _____   K. _____
D. _____   H. _____   L. _____

Copyright by The Goodheart-Willcox Co., Inc.   Modern Welding Lab Workbook   **145**

2. In the GMAW process, changing the wire feed speed also changes the _____.

2. _____

3. Record the proper values or settings for welding mild steel using the GMAW process with the spray transfer method and a .035″ diameter electrode.
   A. Arc voltage range
   B. Amperage range

   3. A. _____
      B. _____

4. Record the proper values or settings for welding aluminum using the GMAW process with the spray transfer method and a 1/16″ diameter electrode.
   A. Arc voltage range
   B. Amperage range

   4. A. _____
      B. _____

5. What is the suggested amperage range when the spray transfer method is used with a 3/32″ diameter electrode on aluminum?

   5. _____

6. Name the parts of the two-drive-roll GMAW wire drive shown in the following image:
   A. _____
   B. _____
   C. _____
   D. _____
   E. _____
   F. _____

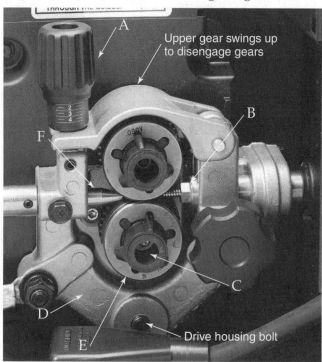

7. The wire rolls shown in the following image are misaligned. Explain what should be moved, and in which direction, to realign these rolls.

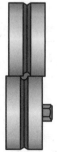

_____

_____

Lesson 9C Gas Metal and Flux Cored Arc Welding—Setting Up the Station

Name _____

8. In the following image, the wire is too loose. What must be done to correct this situation?

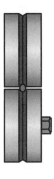

_____
_____
_____

9. The wire and rolls are out of line in the following image. What must be moved, and in what direction, to correct the alignment?

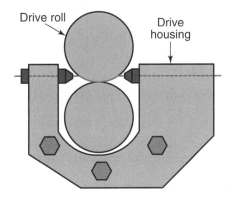

_____
_____

10. Write the letters of the characteristics that apply to argon gas when used in GMAW.
    A. Ionizes easily
    B. Quiets the arc
    C. Produces deep penetration
    D. Requires high voltage
    E. Causes a great deal of spatter
    F. Has low thermal conductivity
    G. Is relatively heavy
    H. Has high thermal conductivity

10. _____

11. Write the letters of the characteristics listed in the previous question that apply to helium gas.

11. _____

12. Which of the following is true of carbon dioxide?
    A. It is the most common shielding gas for FCAW.
    B. It has a lower thermal conductivity than argon.
    C. It is more expensive than argon.
    D. Beads made with carbon dioxide are prone to undercutting.

12. _____

13. Why must the flow rate be increased when welding is performed in the horizontal, vertical, and overhead welding positions using argon and carbon dioxide shielding gases?

_____
_____
_____
_____

14. Name the parts of the nozzle end of the GMAW torch shown in the following image.

    A. _____
    B. _____
    C. _____
    D. _____
    E. _____
    F. _____
    G. _____
    H. _____
    I. _____

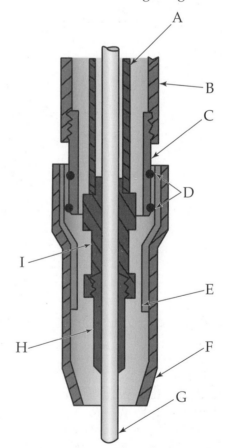

15. What control is used to cause the wire drive motor to slowly feed the welding wire inch by inch through the electrode cable?

_____

Lesson 9C Gas Metal and Flux Cored Arc Welding—Setting Up the Station

Name _____

16. The polarity and gas used in GMAW affects the penetration. Name the gas used and polarity used for each of the welds shown in the following image.

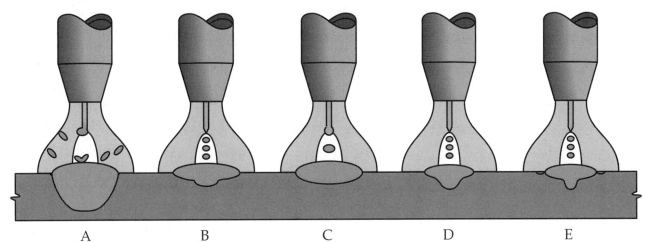

A   B   C   D   E

A. _____
B. _____
C. _____
D. _____
E. _____

17. Name two gases or gas mixtures used to weld stainless steel. See Figures 9-30 and 9-31 in the text.
_____
_____

18. _____ cfh (_____ L/min) is the gas flow rate suggested for welding 18. _____
    3/8″ (9.5mm) thick aluminum. (See Figure 9-34 in the text.)

19. Name the various distances in the following image.

A. _____
B. _____
C. _____
D. _____

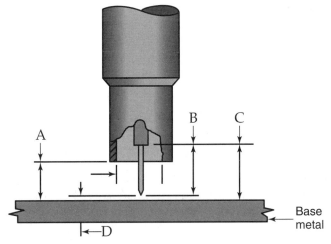

20. *True or False?* A laminar gas flow results when the flow rate is   20. _____
    too high.

Lesson 9D

GMAW and FCAW Procedures for Flat and Horizontal Welds

Name _____ Date _____
Class _____ Instructor _____

Learning Objective
- You will be able to produce welds on lap, inside and outside corners, T-joints, and butt joints in the flat and horizontal welding positions using GMAW or FCAW.

Instructions
Carefully read Headings 9.4 and 9.6 through 9.10 of the text. Also study Figures 9-41 through 9-50 in the text. Then answer the following questions.

1. List two reasons the groove angle used for GMAW or FCAW can be smaller than the angle used for SMAW.
   _____
   _____

2. In GMAW and FCAW, the welder must control what three variables as he or she operates the torch?
   _____
   _____
   _____

Modern Welding Lab Workbook   151

3. Identify which of the following images show the forehand, backhand, and perpendicular methods of welding.

   Backhand welding _____

   Forehand welding _____

   Perpendicular welding _____

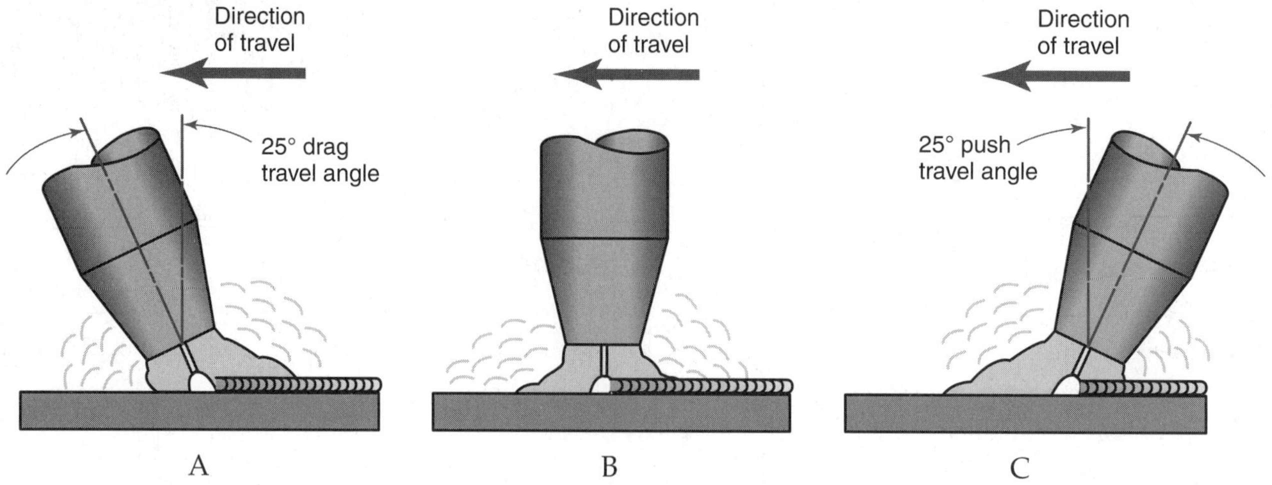

4. Identify the completed welds shown in the following image that generally result from the forehand, backhand, and perpendicular welding methods.

   Backhand welding _____

   Forehand welding _____

   Perpendicular welding _____

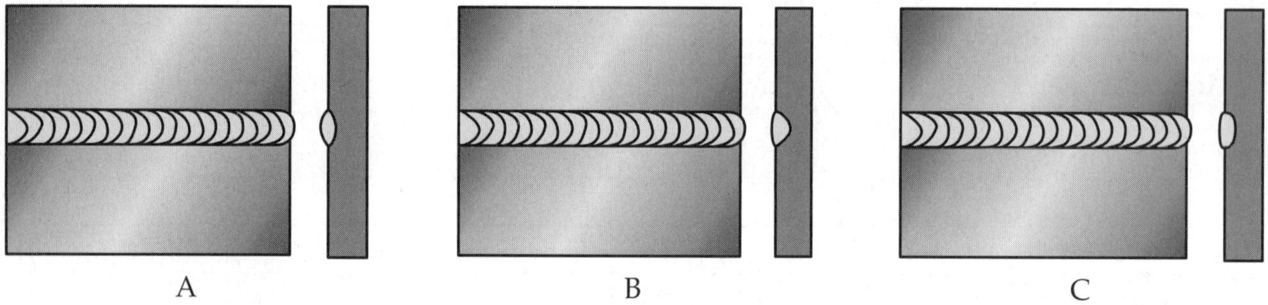

5. In FCAW or GMAW, why is no up-and-down or scratching motion required to strike the arc?

   _____

   _____

   _____

6. *True or False?* A run-off tab can be used to ensure a full-width bead to the end of the weld.   6. _____

152   Modern Welding Lab Workbook

Lesson 9D GMAW and FCAW Procedures for Flat and Horizontal Welds

Name _____

7. The end of a weld should be shielded while it cools by ____.   7. _____
   A. squeezing the trigger on the welding gun
   B. shutting off the ventilation
   C. holding the nozzle at the end of the weld during the preflow of the shielding gas
   D. holding the nozzle at the end of the weld during the shielding gas postflow
   E. reversing the travel of the gun

8. List the six steps required to properly shut down an FCAW or GMAW station.

   _____
   _____
   _____
   _____
   _____
   _____

9. Place the correct electrode angles on the welds shown in the following image. These welds are being made in the flat and horizontal welding positions.

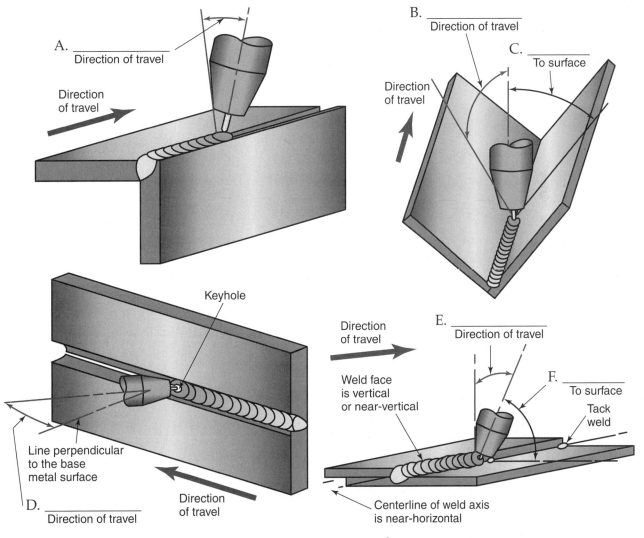

Modern Welding Lab Workbook 153

10. Which metal transfer methods can be used for welding a butt weld in the horizontal position?
   _____
   _____

Job 9D-1

Adjusting GMAW and FCAW Welding Machines

Name _____ Date _____

Class _____ Instructor _____

> **Learning Objective**
>
> • In this job, you will turn on a GMAW or FCAW welding machine. You will also adjust all the variables required to perform gas metal arc welding or flux cored arc welding.

1. Ask your instructor to assign a GMAW machine to you.
2. Study all the controls, switches, and terminals on your assigned machine. Answer the following questions.

    A. Who is the manufacturer of the machine? _____

    B. What is the welding amperage range of the machine? _____

    C. Does the machine have the following? (Circle each correct response.)

    1. Remote control receptacle
    2. Remote contractor receptacle
    3. Pulses per second control
    4. Peak amperage control
    5. Background amperage control
    6. Remote control switch
    7. Remote contactor switch
    8. Ammeter
    9. Voltmeter
    10. Power off/on switch
    11. Wire feed mechanism with a 2- or 4-roll drive system

    D. Does the wire drive or welding machine have the following?

    1. An inch or jog switch
    2. A shielding gas purge switch
    3. A remote control receptacle
    4. A wire feed or retract switch
    5. A wire feed speed control

    E. Which wheel is the driving wheel in each pair? _____

    F. Does your station have the following?

    1. A regulator
    2. A flowmeter

3. Assume that you are welding a butt joint using the GMAW process. You are welding in the flat welding position on an aluminum alloy part that is 1/8" (3.2mm) thick and using the spray transfer method with a 0.047" (1.3mm) electrode. The welding wire should be compatible with the aluminum in the base metal. Answer the following questions.

   A. What arc voltage is suggested? (See Figure 9-17 in the text.)

   B. What amperage range is suggested?

   C. What shielding gas should be used? (See Figure 9-31 in the text.)

   D. What is the suggested shielding gas flow rate? (See Figure 9-34 in the text.)

   3. A. _____

   B. _____

   C. _____

   D. _____

4. Carefully examine and study the wire feed mechanism. Refer to Headings 7.10 and 9.4.2 of the text. Be prepared to explain all the switches, controls, and receptacles on the wire drive mechanism. Also be prepared to show how to correct any drive wheel misalignment on the mechanism.

5. Study your gas cylinder, regulator, and flowmeter. Review Headings 7.3.1 and 7.3.2 of the text regarding cylinder, regulator, and hose safety. Heading 7.3.3 of the text explains the function and operation of the flowmeter. Be prepared to explain and demonstrate for your instructor how the cylinder is opened and how to properly close down the shielding gas part of the GMAW welding station.

6. Study how to remove an empty spool and how to install a full spool of electrode wire.

7. Inform your instructor when you are prepared to locate and explain the functions of all the controls on your welding machine, gas regulator, flowmeter, and wire feed mechanism. Be prepared to explain and demonstrate how to load a spool of electrode wire into the wire feeder.

Instructor's initials: _____

Job 9D-2

Setting a GMAW Welding Machine and Making a Fillet Weld on a Lap Joint in the Flat Welding Position

Name _____ Date _____

Class _____ Instructor _____

> **Learning Objective**
>
> In this job, you will demonstrate your ability to turn on the welding machine and set the variables. You will make a fillet weld on a lap joint in the flat welding position.

1. Obtain four pieces of mild steel that measure 1/16″ × 1 1/2″ × 5″ (1.6mm × 40mm × 125mm).
2. A. Use the short-circuiting transfer method.
   B. Use an ER70S-X or ER80S-X carbon steel welding wire or electrode or see the latest AWS A5.18 specification.
   C. The welding wire diameter should be 0.030″ or 0.035″ (0.8mm or 0.9mm).
   D. Use the shielding gas that is recommended for a short-circuiting transfer on steel less than 1/8″ (3.2mm) with low spatter. Refer to Figures 9-30, 9-32, and 9-34 in the text. What shielding gas is suggested for these conditions?

   _____

3. Before beginning to weld, refer to Figures 9-7, 9-13, 9-30, 9-31, and 9-32 in the text and answer the following questions.

   A. What is the minimum spray arc current or transition current for your electrode?

   _____

   B. Name four shielding gases suggested in the text for use with GMAW on carbon steel.

   _____
   _____
   _____
   _____

   C. The arc voltage should be _____ volts.   C. _____

   D. The suggested amperage range should be between _____   D. _____
      and _____ amperes.

   E. What type current is to be used?   E. _____

See the GMAW gun manufacturer's directions for the correct size of nozzle to use. The general-purpose nozzle size for your gun will usually work well for most applications.

4. Set the voltage, wire feed speed (amperage), and DC polarity on the arc welding machine and wire drive mechanism. Open the shielding gas cylinder. Regulators used with flowmeters are factory set and require no adjustment. The shielding gas flow rates should be about 25 cfh (ft³/hr) or 12 L/min, per Figure 9-34 in the text. When the flow rate is correct, no porosity or discoloring should show in the bead.

5. Make three tack welds in each joint in the weldment shown.

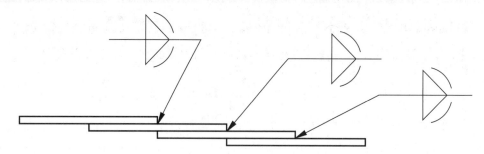

6. Place the weldment into a weld positioning fixture or prop it against firebricks. The welds are to be made in the flat welding position.

7. Complete all fillet welds shown in the drawing in the flat welding position.

8. Inspect each weld and make whatever changes are necessary to improve the next weld made. Increase the gas flow rate if a bead has visible porosity or discoloration.

Inspection

Each weld should be convex in shape. The bead should have an even width with evenly spaced ripples. No porosity or bead discoloration should be visible.

Instructor's initials: _____

Job 9D-3

Fillet Weld on an Inside Corner and a T-Joint in the Flat Welding Position

Name _____ Date _____

Class _____ Instructor _____

> **Learning Objective**
>
> ● In this job, you will weld a fillet weld on an inside corner and a T-joint in the flat welding position.

1. Obtain the following pieces of mild steel:
   1 piece–1/8″ × 4″ × 5″ (3.2mm × 100mm × 125mm)
   1 piece–1/8″ × 1 1/2″ × 4 3/4″ (3.2mm × 40mm × 120mm)
   2 pieces–1/8″ × 1 1/2″ × 4″ (3.2mm × 40mm × 100mm)

2. A. Use the spray arc transfer method.
   B. Use an ER70S-X or ER80S-X carbon steel welding wire or electrode or see the latest AWS A5.18 specification.
   C. The welding wire diameter should be 0.035″ (0.9mm).
   D. Use a gas mixture of Ar + 2%–5% O_2 for these welds.

3. Before beginning to weld, refer to Figures 9-7 and 9-14 in the text and answer the following questions.
   A. What is the minimum spray arc current or transition current for your electrode?
   B. The arc voltage should be _____ volts.
   C. The suggested amperage range should be _____ amperes.
   D. What type current is to be used?

   3. A. _____
      B. _____
      C. _____
      D. _____

See the GMAW gun manufacturer's directions for the correct size of the nozzle to use. The general-purpose nozzle size for your gun will usually work well for most applications.

4. Set the voltage, wire feed speed (amperage), and DC polarity on the arc welding machine and wire drive mechanism. Open the shielding gas cylinder. The shielding gas flow rates should be about 25 cfh (ft³/hr) or 12 L/min. When the flow rate is correct, no porosity or discoloring should show in the bead.

5. Tack weld each joint in three places on the weldment shown.

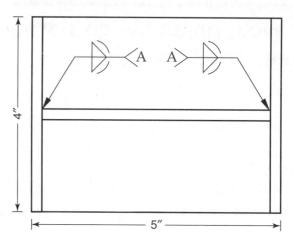

A—¼" Fillet Typical

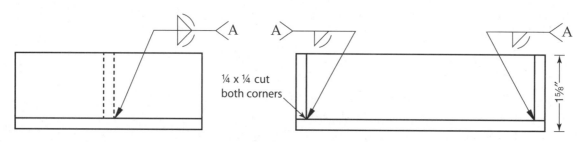

6. Place the weldment into a weld positioning fixture or prop it against firebricks.

7. Complete all welds shown in Step 5 in the flat welding position.

8. Inspect each weld and make whatever changes are necessary to improve the next weld made. Increase the gas flow if the bead has visible porosity or discoloration.

Inspection

Each weld should be convex in shape. The bead should have an even width with evenly spaced ripples. No porosity or bead discoloration should be visible.

Instructor's initials: _____

Job 9D-4

Square-Groove Weld on a Butt Joint in the Flat Welding Position

Name _____ Date _____

Class _____ Instructor _____

> **Learning Objective**
> In this job, you will weld a square-groove butt joint in the flat welding position.

1. Obtain four pieces of mild steel that measure 1/4″ × 1 1/2″ × 5″ (6.4mm × 40mm × 125mm).

2. A. Use the spray arc transfer method.
   B. Use an ER70S-X or ER80S-X carbon steel welding wire, or see the latest AWS A5.18 specification.
   C. Use a gas mixture of Ar + 2%–5% O_2 for these welds.
   D. The welding wire diameter should be 0.045″ (1.1mm).

3. Before beginning to weld, refer to Figures 9-7 and 9-14 in the text and answer the following questions.

   A. What is the minimum spray arc current or transition current for your electrode?
   B. The arc voltage should be _____ volts.
   C. The suggested amperage range should be between _____ and _____ amperes.
   D. What type current is to be used?

   3. A. _____
      B. _____
      C. _____
      D. _____

4. Set the voltage, wire feed speed (amperage), and DC polarity on the arc welding machine or wire drive mechanism. Open the shielding gas cylinder. The shielding gas flow rates should be about 30 cfh (cu.ft/hr) or 14 L/min. See Figure 9-34. When the flow rate is correct, no porosity or discoloring should show in the bead.

5. Grind the required bevel angle(s) on each piece. Tack weld each joint in three places to form the weldment shown.

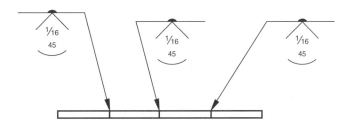

Copyright by The Goodheart-Willcox Co., Inc.

6. Place the weldment into a weld positioning fixture.

7. Complete all welds shown in Step 5 in the flat welding position.

   ■ **Note:** Watch for and maintain a keyhole at the root of the joint. This will ensure 100% penetration.

8. Inspect each weld and make whatever changes are necessary to improve the next weld made. Increase the gas flow rate if the bead has visible porosity or discoloration.

Inspection

Each weld should be convex in shape. The bead should have an even width with evenly spaced ripples. No porosity or bead discoloration should be visible. Complete penetration should be visible on the root side along the entire length of each weld.

Instructor's initials: _____

Job 9D-5

Fillet Weld on a Lap Joint in the Horizontal Welding Position

Name _____ Date _____

Class _____ Instructor _____

> **Learning Objective**
> In this job, you will weld a fillet weld on a lap joint in the horizontal welding position.

1. Obtain four pieces of mild steel that measure 1/8" × 1 1/2" × 5" (3.2mm × 40mm × 125mm).

2. A. Use the spray arc transfer method.
   B. Use an ER70S-X or ER80S-X carbon steel welding wire, or see the latest AWS A5.18 specification.
   C. The welding wire diameter should be 0.035" (0.9mm).
   D. Use a gas mixture of Ar + 2%–5% O_2 for these welds.

3. Before beginning to weld, refer to Figures 9-7 and 9-14 in the text and answer the following questions.

   A. What is the minimum spray arc current or transition current for your electrode?

   B. The arc voltage should be _____ volts.

   C. The suggested amperage range should be _____ to _____ amperes.

   D. What type current is to be used?

   3. A. _____
      B. _____
      C. _____
      D. _____

4. Set the voltage, wire feed speed (amperage), and DC polarity on the arc welding machine or wire drive mechanism. Open the shielding gas cylinder. The shielding gas flow rates should be about 25 cfh (ft³/hr) or 12 L/min. When the flow rate is correct, no porosity or discoloring should show in the bead.

5. Tack weld each joint in three places to form the weldment shown. Tack welds can be made in the flat welding position.

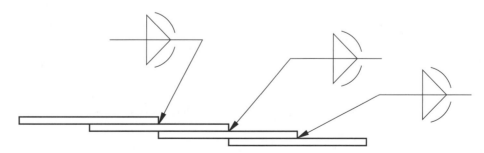

6. Place the weldment into a weld positioning fixture or place it on the welding table. Make all welds in the horizontal welding position.

7. Complete all welds shown in Step 5.

8. Inspect each weld and make whatever changes are necessary to improve the next weld made. Increase the gas flow rate if the bead has visible porosity or discoloration.

Inspection

Each weld should be convex in shape. The bead should have an even width with evenly spaced ripples. No porosity or bead discoloration should be visible.

Instructor's initials: _____

Job 9D-6

Fillet Weld on a T-Joint in the Horizontal Welding Position

Name _____ Date _____

Class _____ Instructor _____

> **Learning Objective**
>
> In this job, you will make a fillet weld on a T-joint in the horizontal welding position.

1. Obtain three pieces of mild steel that measure 1/16″ × 1 1/2″ × 5″ (1.6mm × 40mm × 125mm).
2. A. Use the short-circuiting transfer method.
   B. Use a gas mixture of Ar + 25% CO_2 for these welds.
   C. Use an ER70S-X or ER80S-X carbon steel welding wire, or see the latest AWS A5.18 specification.
   D. The welding wire diameter should be 0.030″ or 0.035″ (0.8mm or 0.9mm).
3. Before beginning to weld, refer to Figure 9-13 in the text and answer the following questions.
   A. The arc voltage should be ____ volts.
   B. The suggested amperage range should be ____ to ____ amperes.
   C. What type current is to be used?

   3. A. _____
      _____
      B. _____
      _____
      C. _____

4. Set the voltage, wire feed speed (amperage), and DC polarity on the arc welding machine or wire drive mechanism. Open the shielding gas cylinder. The shielding gas flow rates should be about 25 cfh (ft^3/hr) or 12 L/min. When the flow rate is correct, no porosity or discoloring should show in the bead.

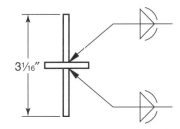

5. Tack weld each joint in three places to form the weldment shown. Tack welds can be made in the flat welding position.

6. Place the weldment into a weld positioning fixture or prop it against firebricks. Make all welds in the horizontal welding position.

7. Complete all welds shown in Step 5.

8. Inspect each weld and make whatever changes are necessary to improve the next weld made. Increase the gas flow rate if the bead has visible porosity or discoloration.

Inspection

Each weld should be convex in shape. The bead should have an even width with evenly spaced ripples. No porosity or bead discoloration should be visible.

Instructor's initials: _____

Lesson 9E

GMAW and FCAW Procedures for Vertical and Overhead Welds

Name _____ Date _____
Class _____ Instructor _____

> **Learning Objective**
>
> ● You will be able to produce welds on lap, inside and outside corners, T-joints, and butt joints in the vertical and overhead positions using GMAW or FCAW. You will also be able to differentiate between semiautomatic and automatic welding and be able to use a troubleshooting guide.
>
> **Instructions**
>
> *Carefully read Headings 9.11 through 9.13 and 9.15 of the text. Also study Figures 9-57 through 9-63 and Figure 9-66 in the text. Then answer the following questions.*

1. Which statement does *not* apply to V-groove butt welds made in the vertical welding position?
   A. The electrode and torch should be inclined (tipped) 20°–25° in the direction of travel.
   B. The electrode centerline should be directly above the weld line.
   C. A push travel angle is used on thin metals.
   D. A keyhole at the root of the weld indicates complete penetration.

   1. _____

2. The preferred GMAW or FCAW metal transfer method(s) for welding in the overhead position is (are) ____.
   A. spray transfer
   B. globular transfer
   C. pulsed spray transfer
   D. short-circuiting transfer

   2. _____

3. Explain why downhill FCAW is difficult to perform. _____
_____
_____

Copyright by The Goodheart-Willcox Co., Inc. Modern Welding Lab Workbook **169**

4. What should you do if the edge melts too quickly while you are making a fillet weld on a lap joint? 4. _____
   A. Increase forward speed.
   B. Point the electrode more toward the surface.
   C. Decrease forward speed.
   D. Increase the wire feed rate.
   E. Decrease the voltage.

5. Why are several narrow beads recommended for overhead welds with FCAW or GMAW?
   _____
   _____
   _____

6. *True or False?* The angle of an electrode is more vertical when a weld is made in the overhead position. 6. _____

7. Study the following image and fill in the answers to questions A and B.
   A. At what angle to the surface is the electrode inclined in the overhead weld? 7. A. _____
   B. At what angle is the electrode tipped in the direction of travel? B. _____

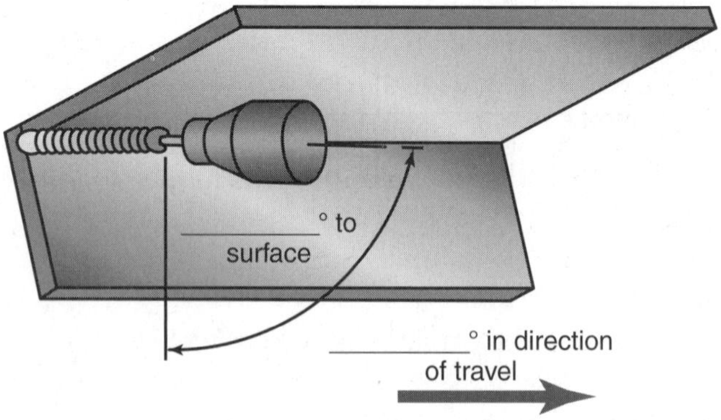

   _____° to surface

   _____° in direction of travel

8. Explain the difference between semiautomatic and automatic welding.
   _____
   _____
   _____
   _____

9. List four causes of undercut on the base metal.
   _____
   _____
   _____

10. How can the condition of excessively wide beads be corrected?
    _____
    _____
    _____

Job 9E-1

Fillet Weld on a Lap Joint in the Vertical Welding Position

Name _____ Date _____

Class _____ Instructor _____

> **Learning Objective**
>
> ● In this job, you will demonstrate your ability to make a fillet weld on a lap joint in the vertical position.

1. Obtain four pieces of mild steel that measure 1/8" × 1 1/2" × 5" (3.2mm × 40mm × 125mm).
2. A. Use the short-circuiting metal transfer method for these welds.
   B. It is recommended that you use an ER70S-X or ER80S-X carbon steel welding wire or electrode or see the latest AWS A5.18 specification.
   C. The welding wire diameter for this job should be 0.030" or 0.035" (0.8mm or 0.9mm).
   D. Use a gas mixture of 75% Ar and 25% CO_2 for these welds.

3. Before you start to weld, refer to Figures 9-7 and 9-13 in the text and answer the following questions.

   A. The arc voltage should be _____ or _____ volts.
   B. The suggested amperage range should be _____ amperes.
   C. What type current is used?

   3. A. _____
      _____
      B. _____
      _____
      C. _____

See the GMAW gun manufacturer's directions for the correct size of the nozzle to use. The general-purpose nozzle size for your gun will usually work well for most applications.

4. Set the voltage, wire feed speed (amperage), and DC polarity on the arc welding machine and wire drive mechanism. Open the shielding gas cylinder. The shielding gas flow rates should be about 25 cfh (ft³/hr) or 12 L/min. When the flow rate is correct, no porosity or discoloring should show in the bead.

5. Make three tack welds in each joint in the weldment shown. Tack weld can be made in the flat position.

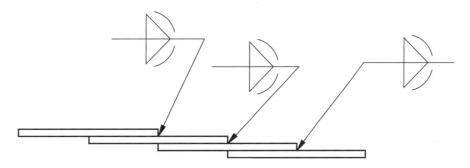

6. Place the weldment into a weld positioning fixture or prop it against firebricks.

7. Complete all welds shown in Step 5 in the vertical welding position.

8. Inspect each weld and make whatever changes are necessary to improve the next weld made. Increase the gas flow rate if the bead has visible porosity or discoloration.

Inspection

Each weld should be convex in shape. The bead should have an even width with evenly spaced ripples. No porosity or bead discoloration should be visible.

Instructor's initials: _____

Job 9E-2

Fillet Weld on an Inside Corner and a T-Joint in the Vertical Welding Position

Name _____ Date _____

Class _____ Instructor _____

> **Learning Objective**
>
> In this job, you will demonstrate your ability to make a fillet weld on an inside corner and T-joint in the vertical position.

1. Obtain eight pieces of mild steel that measure 1/16″ × 1 1/2″ × 5″ (1.6mm × 40mm × 125mm).
2. A. Use the short-circuiting transfer method for these welds.
   B. You should use an ER70S-X or ER80S-X carbon steel electrode wire or welding wire, or refer to the latest AWS A5.18 specification.
   C. The welding wire diameter to use for this job should be 0.030″ or 0.035″ (0.8mm or 0.9mm).
   D. Use a gas mixture of Ar +25% CO_2 for these welds.
3. Before you start to weld, refer to Figure 9-13 in the text and answer the following questions.

   A. The arc voltage should be _____ volts.

   B. The amperage range should be _____ or _____ amperes for the wire diameter you are using.

   C. What type current is to be used?

   3. A. _____

   B. _____
      _____

   C. _____

See the gun manufacturer's recommendation for the correct size nozzle to use. The general-purpose nozzle size for your gun will usually work for most applications.

4. Set the voltage, wire feed speed (amperage), and DC polarity on the arc welding machine and wire drive mechanism. Open the shielding gas cylinder. The shielding gas flow rates should be about 25 cfh (ft³/hr) or 12 L/min. When the flow rate is correct, no porosity or discoloring should show in the bead.
5. Make three tack welds in each joint in the weldment shown. Tack welds can be made in the flat position.

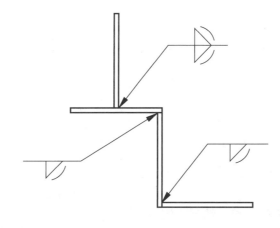

Modern Welding Lab Workbook 173

6. Place the weldment into a weld positioning fixture or prop it against firebricks.

7. Complete all welds shown in Step 5 in the vertical welding position.

8. Repeat Steps 5 and 6 with the remaining pieces of base metal.

9. Inspect each weld and make whatever changes are necessary to improve the next weld made. Increase the gas flow rate if the bead has visible porosity or discoloration.

Inspection

Each weld should be convex in shape. The bead should have an even width with evenly spaced ripples. No porosity or bead discoloration should be visible.

Instructor's initials: _____

Job 9E-3

Bevel-Groove Weld on a Butt Joint in the Vertical Welding Position

Name _____ Date _____

Class _____ Instructor _____

Learning Objective
- In this job, you will weld a bevel-groove butt joint in the vertical position.

1. Obtain five pieces of mild steel that measure 1/4″ × 1 1/2″ × 5″ (6.4mm × 40mm × 125mm).
2. A. Use the short-circuiting transfer method.
   B. Use an ER70S-X or ER80S-X carbon steel electrode wire or refer to the latest AWS A5.18 specification.
   C. The welding wire diameter to use for this job should be 0.035″ (0.9mm)
   D. Use a gas mixture of 75% Ar + 25% CO_2 for these welds.
3. Before you start to weld, refer to Figure 9-13 in the text and answer the following questions.
   A. The peak arc voltage should be _____ volts.
   B. The suggested peak amperage range should be _____ amperes.
   C. What type current is to be used?

   3. A. _____
      B. _____
      C. _____

See the gun manufacturer's recommendation for the correct size nozzle to use. The general-purpose nozzle size for your gun will usually work for most applications.

4. Set the voltage, wire feed speed (amperage), and DC polarity on the arc welding machine and wire drive mechanism. Open the shielding gas cylinder. The shielding gas flow rates should be about 25 cfh (ft³/hr) or 12 L/min. When the flow rate is correct, no porosity or discoloring should show in the bead.
5. Study the following drawing and determine which pieces require a beveled edge. Grind a bevel on the required pieces. Make three tack welds in each joint in the weldment shown. Tack welds can be made in the flat position.

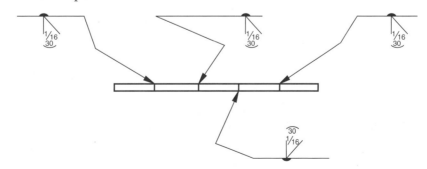

6. Place the weldment into a weld positioning fixture or prop it against firebricks so that all welds are made in the vertical position.

7. Complete all welds shown in Step 5.

   ■ **Note:** Watch for and maintain a keyhole at the root of the joint. This will ensure 100% penetration.

8. Inspect each weld and make whatever changes are necessary to improve the next weld made. Increase the gas flow rate if the bead has visible porosity or discoloration.

Inspection

Each weld should be convex in shape. The bead should have an even width with evenly spaced ripples and uniform penetration. No porosity or bead discoloration should be visible.

Instructor's initials: _____

Job 9E-4

Fillet Weld on a Lap Joint in the Overhead Welding Position

Name _____   Date _____

Class _____   Instructor _____

> **Learning Objective**
> In this job, you will produce fillet welds on a lap joint in the overhead welding position.

> **Caution:** Since this weld is made above your head, you must wear a leather cape or jacket on your shoulders, leather spats, and a cap on your head to cover your hair. Also observe all other arc welding safety precautions.

1. Obtain four pieces of mild steel that measure 1/16″ × 1 1/2″ × 5″ (1.6mm × 40mm × 125mm).
2. A. Use the short-circuiting transfer method.
   B. Use an ER70S-X or ER80S-X carbon steel electrode wire or refer to the latest AWS A5.18 specification.
   C. The welding wire diameter to use for this job should be .035″ (0.9mm).
   D. What shielding gas mixture should be used?
   _____
3. Before you start to weld, refer to Figure 9-13 in the text and answer the following questions.
   A. The arc voltage should be _____ volts.   3. A. _____
   B. The suggested amperage range should be _____ or _____   B. _____
   amperes.   C. _____
   C. What type current is to be used?

See the gun manufacturer's recommendation for the correct size nozzle to use. The general-purpose nozzle size for your gun will usually work for most applications.

4. Set the voltage, wire feed speed (amperage), and DC polarity on the arc welding machine and wire drive mechanism. Open the shielding gas cylinder. The shielding gas flow rates should be about 25 cfh (ft³/hr) or 12 L/min. The flow rate is slightly higher when welding overhead. When the flow rate is correct, no porosity or discoloring should show in the bead.
5. Make three tack welds in each joint in the weldment shown. Tack welds can be made in the flat position.

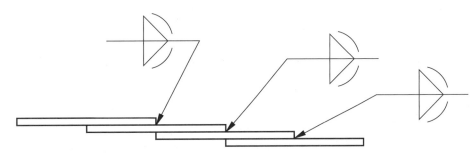

Modern Welding Lab Workbook   **177**

6. Place the weldment into a weld positioning fixture.

7. Complete all welds shown in Step 5 in the overhead welding position.

8. Inspect each weld and make whatever changes are necessary to improve the next weld made. Increase the gas flow rate if the bead has visible porosity or discoloration.

Inspection

Each weld should be convex in shape. The bead should have an even width with evenly spaced ripples. No porosity or bead discoloration should be visible.

Instructor's initials: _____

Job 9E-5

Fillet Weld on a T-Joint in the Overhead Welding Position

Name _____ Date _____

Class _____ Instructor _____

Learning Objective
- In this job, you will produce a fillet weld on a T-joint in the overhead welding position.

■ **Caution:** Since this weld is made above your head, you must wear a leather cape or jacket on your shoulders, leather spats, and a cap on your head to cover your hair. Also observe all other arc welding safety precautions.

1. Obtain three pieces of mild steel that measure 1/8" × 1 1/2" × 5" (3.2mm × 40mm × 125mm).

2. A. Use the pulsed spray transfer method for these welds. If your machine does not have pulsed spray, use the short-circuiting transfer method.

   B. Use an ER70S-X or ER80S-X carbon steel electrode wire or welding wire, or refer to the latest AWS A5.18 specification.

   C. The welding wire diameter to use for this job should be 0.035" (0.9mm).

   D. What shielding gas mixture should be used?
   _____

3. Before you start to weld, refer to Figures 9-7 and 9-13 or 9-14 in the text and answer the following questions.

   A. What is your electrode's minimum transition current?   3. A. _____

   B. The peak arc voltage should be ____ volts.   B. _____

   C. The suggested peak amperage range should be between   C. _____
      ____ and ____ amperes.

   ■ **Note:** Follow the recommendation for the power supply to set up the pulsed spray transfer. The background current is set to about one-half the peak current value. Some machines set a percentage for the background current. Start with 50%. See Heading 9.2.4 and Figure 9-8.

   D. What type current is to be used?
   _____

See the gun manufacturer's recommendation for the correct size nozzle to use. The general-purpose nozzle size for your gun will usually work for most applications.

4. Set the voltage, wire feed speed (amperage), and DC polarity on the arc welding machine and wire drive mechanism. Open the shielding gas cylinder. The shielding gas flow rates should be about 25 cfh (ft³/hr) or 12 L/min. When the flow rate is correct, no porosity or discoloring should show in the bead.

Copyright by The Goodheart-Willcox Co., Inc.   Modern Welding Lab Workbook   **179**

5. Make three tack welds in each joint in the weldment shown. Tack welds can be made in the flat position.

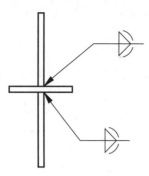

6. Place the weldment into a weld positioning fixture.

7. Complete all welds shown in Step 5 in the overhead welding position.

8. Inspect each weld and make whatever changes are necessary to improve the next weld made. Increase the gas flow rate if the bead has visible porosity or discoloration.

Inspection

Each weld should be convex in shape. The bead should have an even width with evenly spaced ripples. No porosity or bead discoloration should be visible.

Instructor's initials: _____

Job 9E-6

Square-Groove Weld on a Butt Joint in the Overhead Welding Position

Name _____ Date _____

Class _____ Instructor _____

Learning Objective
- In this job, you will make a square-groove weld on a butt joint in the overhead welding position.

■ **Caution:** Since you are making this weld above your head, you must wear a leather cape or jacket on your shoulders, leather spats, and a cap on your head to cover your hair. Also, observe all other arc welding safety precautions.

1. Obtain four pieces of mild steel that measure 1/16″ × 1 1/2″ × 5″ (1.6mm × 40mm × 125mm).
2. A. Use the short-circuiting transfer method.
   B. Use an ER70S-X or ER80S-X carbon steel electrode wire or welding wire, or refer to the latest AWS A5.18 specification.
   C. The welding wire diameter to use for this job should be 0.035″ (0.9mm).
   D. Use a gas mixture of Ar + 25% CO_2 for these welds.
3. Before you start to weld, refer to Figure 9-13 in the text and answer the following questions.
   A. The arc voltage should be ____ volts.   3. A. _____
   B. The suggested amperage range is ____ amperes.   B. _____
   C. What type current is to be used?   C. _____

See the gun manufacturer's recommendation for the correct size nozzle to use. The general-purpose nozzle size for your gun will usually work for most applications.

4. Set the voltage, wire feed speed (amperage), and DC polarity on the arc welding machine and wire drive mechanism. Open the shielding gas cylinder. The shielding gas flow rates should be about 25 cfh (ft³/hr) or 12 L/min. When the flow rate is correct, no porosity or discoloring should show in the bead.
5. Make three tack welds in each joint in the weldment shown. Tack welds can be made in the flat position.

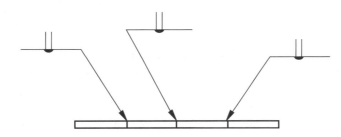

6. Place the weldment into a weld positioning fixture so that all welds are made in the overhead position.

7. Complete all welds shown in Step 5.

   ■ **Note:** Watch for and maintain a keyhole at the root of the joint. This will ensure 100% penetration.

8. Inspect each weld and make whatever changes are necessary to improve the next weld made. Increase the gas flow rate if the bead has visible porosity or discoloration.

Inspection

Each weld should be convex in shape. The bead should have an even width with evenly spaced ripples and uniform penetration. No porosity or bead discoloration should be visible.

Instructor's initials: _____

Lesson 10
Plasma Arc Cutting

Name _____ Date _____
Class _____ Instructor _____

> **Learning Objective**
>
> • You will be able to identify equipment and supplies used in the plasma arc cutting and gouging process. You will also be able to describe the principles used in this process.
>
> **Instructions**
>
> *Carefully read Headings 10.1 through 10.8 of the text and study the related figures. Then answer the following questions.*

1. Free electrons in plasma conduct _____. 1. _____

2. Plasma arc welding power supplies are constant _____ machines. 2. _____

3. Label the parts of the plasma arc cutting torch shown.

   A. _____
   B. _____
   C. _____
   D. _____
   E. _____
   F. _____

4. List two ways cutting torches can be cooled. _____

5. Which of the following is used as an electrode material? 5. _____
   A. Mild steel
   B. Stainless steel
   C. Aluminum
   D. Hafnium

Copyright by The Goodheart-Willcox Co., Inc.   Modern Welding Lab Workbook   183

6. The plasma gas and arc travel through a hole in which part of the torch? _____
_____
_____

7. What polarity is the electrode in plasma arc cutting?   7. _____

8. Name the two purposes of gases used in PAC. _____
_____

9. The most common gas used for manual plasma arc cutting is _____.   9. _____

10. What plasma gas is recommended for cutting aluminum that is 1.0" (25mm) thick? _____
_____

11. What minimum number lens should be worn during plasma arc cutting?   11. _____

12. What can be worn to protect a person from the loud noises that can occur during plasma arc cutting?   12. _____

13. *True or False?* If the proper current setting is not known, it is better to set the current lower than needed prior to cutting.   13. _____

14. A pilot arc is between the _____ and the _____.   14. _____

15. Which type of arc erodes the electrode and constricting nozzle more—a transferred arc or nontransferred arc?   15. _____

16. Describe how to pierce using the PAC process. _____
_____
_____
_____

17. What causes dross that is difficult to remove from the base metal? _____
_____

18. During plasma arc cutting, a proper cutting speed is indicated when sparks exit from the bottom of the base metal and point rearward by about _____°.   18. _____

19. Plasma arc gouging requires _____.   19. _____
    A. a greater voltage than plasma arc cutting
    B. a smaller constricting nozzle orifice than plasma arc cutting
    C. the torch to be held perpendicular to the base metal
    D. All of the above.

20. List three shielding gases used for plasma arc gouging.
_____

Job 10-1

Piercing and Cutting Using the Plasma Arc Cutting Process

Name _____ Date _____

Class _____ Instructor _____

> **Learning Objective**
> - In this job, you will pierce and cut various materials using the PAC process.

1. Obtain pieces of 1/4" (6.4mm) mild steel, stainless steel, and aluminum that are 3" (75mm) wide and 10" (250mm) long.
2. Refer to Figures 10-11, 10-15, 10-16, and 10-17 in the text and answer the following questions.

   To cut the mild steel:

   A. What polarity is suggested for cutting or piercing? _____

   B. What current setting is suggested? _____

   C. What low-cost plasma gas will produce a good-quality cut? _____

   D. What is the recommended cutting speed? _____

   To cut the stainless steel:

   A. What polarity is suggested for cutting or piercing? _____

   B. What current setting is suggested? _____

   C. What low-cost plasma gas will produce a good-quality cut? _____

   D. What is the recommended cutting speed? _____

   To cut the aluminum:

   A. What polarity is suggested for cutting or piercing? _____

   B. What current setting is suggested? _____

   C. What low-cost plasma gas will produce a good-quality cut? _____

   D. What is the recommended cutting speed? _____

3. Mark each piece of metal as shown on the following drawing.

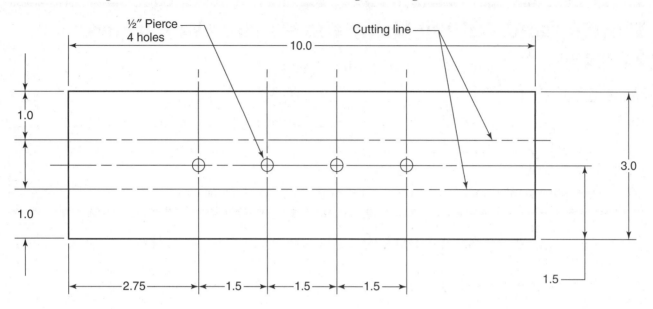

4. Pierce four holes approximately 1/2" (13mm) in diameter along the line drawn in the middle of each piece.

5. Cut the 3" (75mm) pieces in two places as shown on the drawing.

Inspection

The edges of the holes and the edge of the cut metal should be smooth. The top and bottom edges should be free of dross.

Instructor's initials: _____

Job 10-2

Cutting a Shape Using the Plasma Arc Cutting Process

Name _____ Date _____
Class _____ Instructor _____

Learning Objective
- In this job, you will cut a shape on plain carbon steel, stainless steel, and aluminum using the PAC process.

1. Obtain pieces of 1/4" (6.4mm) mild steel, stainless steel, and aluminum that are 2" (50mm) wide and 6" (150mm) long.

2. Using soapstone, draw the shape shown in the following diagram on each piece of base metal.

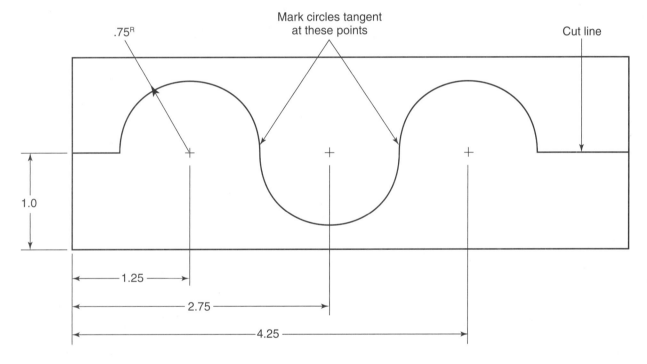

3. Using the variables listed in Job 10-1, cut along the lines drawn on each piece of base metal.

4. What action needs to be taken if there is hard-to-remove dross on the bottom edge of the metal?

Inspection
The edges of the shape you cut should be square and fairly smooth. The surfaces should be free of dross. The shape should be cut as close to the drawn line as possible.

Instructor's initials: _____

Modern Welding Lab Workbook **187**

Lesson 11A

Oxyfuel Gas Welding Equipment and Supplies

Name _____   Date _____

Class _____   Instructor _____

Learning Objective

You will be able to identify the parts and equipment included in an oxyfuel gas welding station. You will be able to describe the function and assembly of the parts. You will also be able to list the supplies needed for oxyfuel gas welding.

Instructions

Carefully read Headings 11.1 through 11.11.8 of the text. Also study Figures 11-1 through 11-73 in the text. Then answer the following questions.

1. Label the parts of the oxyacetylene welding outfit shown.

   A. _____
   B. _____
   C. _____
   D. _____
   E. _____
   F. _____
   G. _____
   H. _____
   I. _____
   J. _____
   K. _____
   L. _____
   M. _____
   N. _____

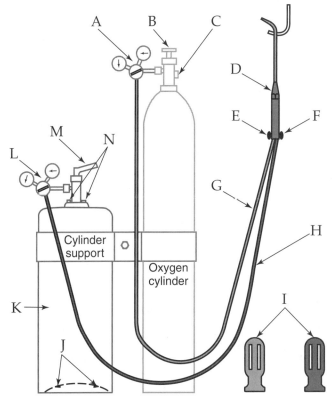

Copyright by The Goodheart-Willcox Co., Inc.   Modern Welding Lab Workbook   189

2. Liquid oxygen is generally stored and shipped in a(n) _____ flask.

2. _____

3. Oxygen cylinder construction specifications are prepared by the _____. No part of the cylinder is less than _____ inch thick. The cylinder must withstand _____ psig of hydrostatic pressure.

3. _____
_____
_____

4. New, empty acetylene cylinders are filled with an inert, _____ (massively uniform) filler material that cures to a porosity of 85%.

4. _____

5. The rate at which acetylene can be drawn off or removed from an acetylene cylinder depends on the _____.
   A. amount of charge remaining in the cylinder
   B. temperature of the cylinder
   C. number of cylinders providing the flow
   D. All of the above.
   E. None of the above.

5. _____

6. Mixing _____ and water will produce acetylene gas.

6. _____

7. Oxygen is stored in cylinders at pressures from 2000 psig to 2600 psig (14000kPa to 18000kPa). Welding torches operate at an oxygen pressure of _____ psig. Therefore, a pressure regulator is required to reduce and control the pressure.
   A. 3 to 50
   B. 1 to 30
   C. 0 to 30
   D. 0 to 50
   E. 1 to 50

7. _____

8. Name the parts shown in the schematic of a single-stage nozzle-type pressure regulator.
   A. _____
   B. _____
   C. _____
   D. _____
   E. _____
   F. _____
   G. _____
   H. _____
   I. _____
   J. _____
   K. _____
   L. _____
   M. _____

Lesson 11A Oxyfuel Gas Welding Equipment and Supplies

Name _____

9. A two-stage regulator reduces the cylinder pressure to a working pressure in two stages. The lowest first stage pressure is generally _____ kPa (metric) of pressure.
   A. 200
   B. 34.5
   C. 5
   D. 1379
   E. 2000

9. _____

10. A regulator is turned completely off when the regulator screw is turned all the way _____.

10. _____

11. Why is the area above 15 psig marked with a red band on some acetylene regulators? _____
_____
_____

12. Hoses are made in three colors: _____, _____, and _____. _____
Which color is usually used for the fuel gas hose? _____

13. The acetylene or fuel gas hose nut is different from the oxygen hose nut. The acetylene nut has a(n) _____ cut around it, has left-hand threads, and may have _____ stamped on it.

13. _____
    _____

14. Label the parts of the positive-pressure oxyacetylene torch shown in the following image.

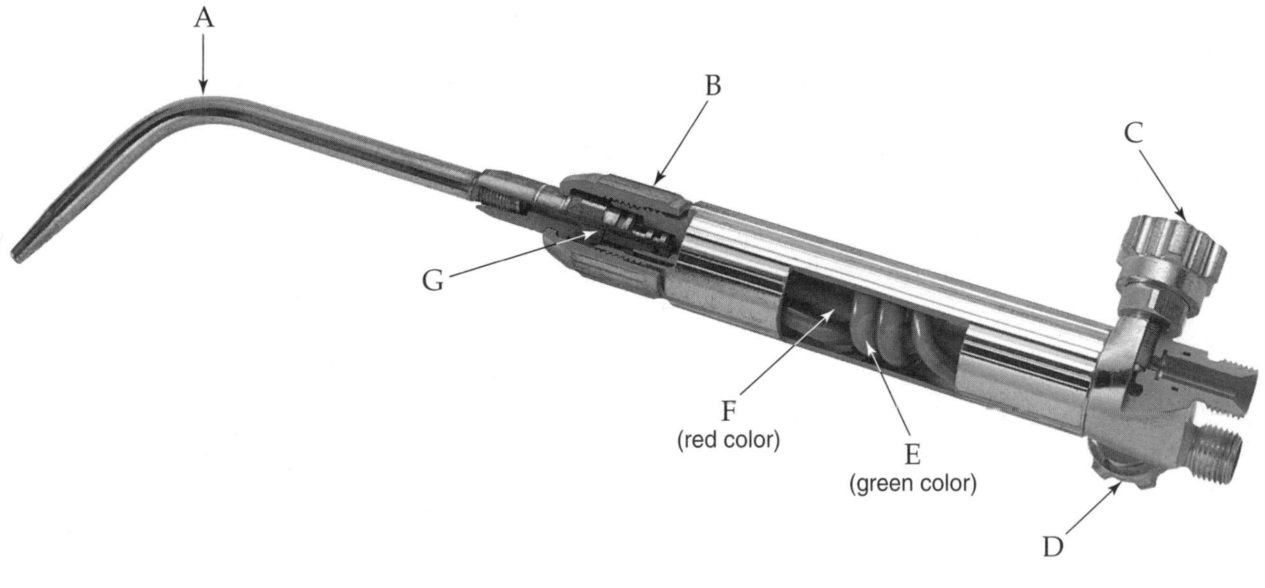

A. _____   E. _____
B. _____   F. _____
C. _____   G. _____
D. _____

15. Special welding tip cleaners consist of a series of _____ wires that correspond to the diameters of tip orifices.

15. _____

16. List the recommended welding lens shade number for use in each of the following welding or cutting processes.

    A. Medium gas welding

    B. Heavy gas welding

    C. Light gas welding

    16. A. _____

    B. _____

    C. _____

17. Flint and steel _____ lighters are generally used to light the oxyfuel gas welding or cutting flame.

    17. _____

18. List five gases used with the oxyfuel gas process. _____

    _____

    _____

    _____

19. Hoses should never be interchanged (switched). If _____ is passed through a hose previously used for acetylene, a combustible mixture may form and cause a flame in the hose.

    19. _____

20. How do flashback arrestors prevent flashbacks from occurring? _____

    _____

    _____

21. Name five common metals used to make welding rods. _____

    _____

    _____

    _____

    _____

    _____

22. Most welding rods are _____" long. They are packaged in _____ lb (_____ kg) bundles.

    22. _____

    _____

    _____

23. Welding rods made for oxyacetylene welding are the RG45, RG60, and RG65. The tensile strength of the RG60 welding rod is _____ psi or _____ Mpa.

    23. _____

    _____

Lesson 11B

Oxyfuel Gas Welding and Cutting Safety

Name _____ Date _____

Class _____ Instructor _____

Learning Objective
- You will be able to safely use oxyfuel gas welding equipment.

Instructions
Carefully read Headings 11.2 through 11.11.1, 12.2.1 through 12.2.5, 12.5.3, 12.7, 12.8 through 12.8.7, and 14.10 of the text. Also study the related figures in the text. Then answer the following questions.

1. The oxygen cylinder valve is constructed using a back-seating valve to seal the stem from leakage. Therefore, when opening the oxygen cylinder, the valve must be ____.
   A. fully closed
   B. half way open
   C. fully opened
   D. at the correct pressure position

   1. _____

2. *True or False?* When a cylinder is not in use, the cylinder valve must be closed whether the cylinder is full or empty.

   2. _____

3. Liquid oxygen is stored at ____°F (____°C). It may cause freeze burns on the eyes or skin if it comes in contact with them.

   3. _____

4. List seven organic materials that must be kept away from oxygen to prevent fires from occurring.
   _____
   _____
   _____

5. Clothing that has been saturated by oxygen should be removed and not worn again for at least ____ minutes, or until no excess oxygen remains in it.

   5. _____

Modern Welding Lab Workbook

6. Why is it recommended that you do not carry matches and other combustible items in your pockets?
   _____
   _____
   _____

7. *True or False?* The storage of acetylene (C_2H_2) in its gaseous form is not permitted at pressures above 15 psig (103kPa).

7. _____

8. Fuse plugs, which permit acetylene to be released from the acetylene cylinder in case of a fire, will melt at _____°F (_____°C).

8. _____
   _____

9. *True or False?* A lighted torch should never be aimed toward the cylinders or hoses.

9. _____

10. Piping made of _____ must never be used in the presence of acetylene.

10. _____

11. Which statement does not apply to acetylene and acetylene containers?
    A. Acetylene smells like garlic.
    B. Too much acetylene in the air can cause dizziness.
    C. Always leave a cylinder valve wrench, wheel, or key in place for emergency shutoff.
    D. Acetylene is stored at 2000 psig (13790kPa) pressure.
    E. Flames should be kept away from cylinder fuse plugs.

11. _____

12. Oxygen and acetylene cylinder valves are closed when they are turned all the way _____.

12. _____

13. Oxygen and acetylene regulators are closed when the regulator adjusting screw is turned all the way _____.

13. _____

14. The filter lenses worn during oxyfuel gas cutting or welding protect the eyes from _____ and _____ rays.

14. _____

15. *True or False?* Fuel cylinder valve threads are usually left-hand threads. Oxygen cylinder valves have right-hand threads. The thread diameters are also different. This is done to prevent connecting the wrong regulator to a cylinder.

15. _____

16. Cylinder valves are protected from damage, while they are stored or moved, by threading on a properly fitting _____.

16. _____

17. Cylinders should be moved using a(n) _____.

17. _____

18. If a flashback occurs, first close the torch _____ valve, and then close the _____ torch valve.

18. _____
    _____

194   Modern Welding Lab Workbook

Lesson 11B Oxyfuel Gas Welding and Cutting Safety

Name _____

19. Why should you lay the welding rod down with the hot end away from your body when repositioning your grip on the rod?

    _____

    _____

20. List the four types of clothing a welder should wear when welding in the overhead position.

    _____

    _____

    _____

21. List four metals that release toxic fumes.

    _____

    _____

22. The flame should always be _____ when the torch is not in your hand.   22. _____

23. Why should a welder never stand in front of the regulator gauges when the cylinder is being turned on?

    _____

    _____

    _____

24. Tanks and containers that hold flammable or explosive materials should be welded only under the supervision of a qualified _____ engineer.   24. _____

25. Backfires are generally caused by what three conditions?

    _____

Lesson 12A

Oxyfuel Gas Welding—Turning the Outfit On and Off

Name _____ Date _____

Class _____ Instructor _____

Learning Objective

You will be able to turn on an oxyacetylene welding outfit, light and adjust a positive-pressure type torch, and shut down an oxyacetylene welding outfit.

Instructions

Carefully read the Procedure for Turning on the Oxyacetylene Outfit and Lighting a Positive-Pressure Type Torch section under Heading 12.2 of the text. Also read Headings 12.1.3, 12.2, 12.2.1, 12.2.4, 12.2.5, and 12.2.6 and study Figures 12-3 and 12-13 through 12-15. Then answer the following questions.

1. *True or False?* The high-pressure and low-pressure gauges on both the acetylene and oxygen regulators should read 3–5 psig after the oxyacetylene outfit has been properly shut down.

   1. _____

2. *True or False?* Before any welding equipment is used, it is essential to check the condition of the equipment to make sure the welding outfit is properly assembled.

   2. _____

Refer to the following illustration showing carburizing, oxidizing, and neutral oxyacetylene welding flames to answer Questions 3–5.

3. Which of the flames shown is neutral?

   3. _____

4. Which of the flames shown is oxidizing?

   4. _____

5. Which of the flames shown is carburizing?

   5. _____

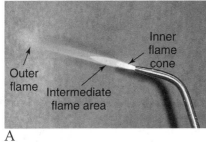
A

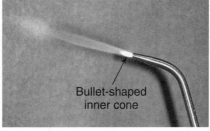
B

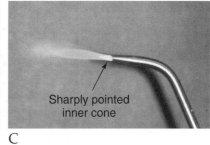
C

6. The following are the steps required to turn on an oxyacetylene welding outfit and light a positive-pressure type torch. They are not in the correct order. Place them in order by writing the correct letter in the appropriate blank.

   ■ **Note:** In the correct order, the first eight steps will relate to turning on the outfit, and the last three steps will relate to lighting the torch.

   A. Open the oxygen cylinder valve.
   B. Open the torch oxygen valve one turn. Turn the oxygen regulator adjusting screw in (clockwise) until the low-pressure oxygen gauge indicates the working pressure that is correct for the tip orifice. Then, turn off the oxygen torch valve.
   C. Visually check the condition of the equipment.
   D. Open the acetylene torch valve no more than 1/16 turn. Use a spark lighter to ignite the acetylene gas coming out of the tip.
   E. Slowly open the acetylene cylinder valve 1/4 turn to 1/2 turn counterclockwise.
   F. Turn the regulator adjusting screws all the way out (counterclockwise) before opening the cylinder valves.
   G. Slowly open the oxygen valve on the torch.
   H. Open the acetylene torch valve one turn. Turn the acetylene regulator adjusting screw in slowly (clockwise) until the low-pressure acetylene gauge indicates a pressure that is correct for the tip size. After setting the acetylene working pressure, turn off the acetylene torch valve using finger-tip force only.
   I. Inspect the regulators.
   J. Continue to open the acetylene torch valve slowly until the acetylene flame jumps away from the end of the tip slightly and no longer smokes.
   K. Stand to one side of the regulator when opening the cylinder valves. A bursting regulator or gauge could cause severe injury.

   1. _____
   2. _____
   3. _____
   4. _____
   5. _____
   6. _____
   7. _____
   8. _____
   9. _____
   10. _____
   11. _____

7. The steps for shutting down an oxyacetylene outfit are as follows. Place them in order by writing the correct letter in the appropriate blank.

   A. Wait until the high-pressure and the low-pressure gauges on both the acetylene and the oxygen regulators read zero.
   B. Close the acetylene torch valve, then the oxygen torch valve. This extinguishes the flame and eliminates the soot.
   C. Open the hand valves on the torch.
   D. Turn the adjusting screws on both the acetylene and oxygen regulators all the way out.
   E. Lightly close both hand valves on the torch, then hang up the torch.
   F. Tightly close the cylinder valves.

   1. _____
   2. _____
   3. _____
   4. _____
   5. _____
   6. _____

Lesson 12A Oxyfuel Gas Welding—Turning the Outfit On and Off

Name _____

8. If the low-pressure gauge reading continues to creep upward after the torch valve is turned off, this indicates a leaking regulator nozzle and seat. If this happens, turn off the _____ valve immediately, or the low-pressure gauge may burst.
   A. regulator
   B. torch
   C. cylinder
   D. gauge
   E. station

8. _____

9. The oxyfuel gas flame should be lighted using a(n) _____.

9. _____

10. *True or False?* The torch should not be pointed at another person while it is being lighted.

10. _____

Job 12A-1

Turning On, Lighting, and Shutting Down an Oxyacetylene Welding Outfit

Name _____ Date _____

Class _____ Instructor _____

> **Learning Objective**
>
> ● In this job, you will safely and properly turn on an oxyacetylene welding outfit equipped with a positive-pressure torch. You will also light and adjust the flame and correctly turn the outfit off. You will use the steps listed in the "Procedure for Turning on the Oxyacetylene Outfit and Lighting a Positive-Pressure Type Torch" under Heading 12.2 in the text. Also refer to the steps listed in "Procedure for Shutting Down an Oxyacetylene Outfit" under Heading 12.2.6 in the text.

Turning on the Oxyacetylene Welding Outfit

1. Work with a group of two or three students on an assigned oxyacetylene welding outfit.

   ■ **Note:** When setting the oxygen and acetylene pressures and selecting the torch tip size, assume that you will be using a positive-pressure welding torch. Assume also that you will be welding on 1/8″ (3.2mm) base metal.

2. Determine the manufacturer of the tip. What is the manufacturer's name? _____
   _____

3. Determine the correct tip orifice drill size to be used for a metal thickness of 1/8″ (3.2mm) steel. See Figure 12-13 in the text.

   What is the size? _____

4. Determine the correct acetylene and oxygen pressures from the table in Figure 12-13 of the text. What are the pressures here in lbs/in^2 gauge (psig)?

   Acetylene _____ psig

   Oxygen _____ psig

5. The first student will tell the other member(s) of the group what is to be done to turn on the station, starting from the first step.

6. All members must agree that this is the correct thing to do and is in the correct order. After they agree, the first student will do what is agreed upon by the group.

7. The first student will proceed through each step, waiting until everyone agrees. Stop when Step 8 in "Procedure for Turning on the Oxyacetylene Outfit and Lighting a Positive-Pressure Type Torch" is reached.

8. Have your instructor approve your work.

Instructor's initials: _____

Lighting and Adjusting the Oxyacetylene Flame

1. The first student will continue on to light the flame and adjust it to neutral.

2. Before doing any step, the first student must tell the group what he or she intends to do. They all must agree before the step is performed.

3. The first student will continue step-by-step until the torch is lighted and adjusted to a neutral flame.

4. Have your instructor approve your work.

Instructor's initials: _____

Shutting Down the Oxyacetylene Outfit

1. The first student will proceed to shut down the outfit.

2. Before doing any step, the first student must tell the group what he or she intends to do. They must all agree before the step is performed.

3. The first student will continue through the proper steps until the welding outfit is completely and correctly shut down.

Repeat the Procedures

Each student will go through the entire turning on, lighting and adjusting, and shutting down procedure in the same manner as the first student.

■ **Caution:** Be careful not to trap pressure in the high-pressure side of the regulator and in the high-pressure gauge. This will indicate an incorrect shut-down procedure.

Instructor's initials: _____

Lesson 12B

Oxyfuel Gas Welding—Running a Continuous Weld Pool

Name _____ Date _____

Class _____ Instructor _____

Learning Objective
- You will be able to select the correct torch tip and oxygen and acetylene gas pressures. You will be able to run a continuous weld pool on mild steel. You will also be able to make an outside corner joint without a welding rod.

Instructions
Carefully read Headings 12.2.3, 12.2.4, and 12.3 through 12.3.6 of the text. Also study the tables in Figures 12-13 and 12-16 in the text. Then answer the following questions.

1. What should you do if the low-pressure gauge needle continues to rise or creep upward after the torch valve is turned off?

   _____

   _____

2. To weld a 1/16″ piece of low carbon steel, with a positive-pressure type welding torch, you should use a # _____ drill size tip, _____ psig of oxygen, and _____ psig acetylene.

   2. _____

3. To weld a 3/32″ thick piece of metal with a positive-pressure torch, you should use a # _____ drill size tip, _____ psig of oxygen, and _____ psig of acetylene.

   3. _____

4. To weld a 1/8″ thick piece of metal with an injector-type torch, you should use a # _____ drill size tip, _____ psig of oxygen, and _____ psig of acetylene.

   4. _____

5. When running a continuous weld pool, the torch tip should be held at a _____ travel angle.
   A. 15°–30°
   B. 20°–30°
   C. 25°–45°
   D. 35°–45°
   E. 60°–70°

   5. _____

Copyright by The Goodheart-Willcox Co., Inc.

6. In the space provided, draw three different torch motions that can be used when welding or running a continuous weld pool.

```
   ┌─────────────────────────────┐
A  │    ———————  - -  ———————    │
   │                             │
B  │    ———————  - -  ———————    │
   │                             │
C  │    ———————  - -  ———————    │
   └─────────────────────────────┘
```

7. The tip of the inner flame should be _____ from the base metal surface.
   A. 1/32"–1/8" (0.8mm–1.6mm)
   B. 1/16"–1/8" (1.6mm–3.2mm)
   C. 1/8"–1/4" (3.2mm–6.4mm)

   7. _____

8. List four things an experienced welder can determine by watching the weld pool.

   _____

   _____

   _____

   _____

9. List three things that can be done with the torch tip to reduce the heat and decrease the pool size if the weld pool gets too wide.

   _____

   _____

   _____

10. A well-made continuous weld pool will be straight, with even width, no holes, and good _____.

    10. _____

Job 12B-1

Running a Continuous Weld Pool

Name _____ Date _____

Class _____ Instructor _____

> **Learning Objective**
>
> • In this job, you will run a continuous weld pool. This exercise teaches you how to control the molten metal without melting through to the other side.

1. Obtain a piece of mild steel that measures 1/16″ × 3″ × 5″ (1.6mm × 75mm × 125mm).

2. Draw five evenly spaced lines on the metal surface using chalk or soapstone. These lines must run along the 5″ (125mm) length, one in the middle and two to each side. The outer lines should be 3/8″ (9.5mm) from the edges.

3. What type welding torch is used on your welding outfit?   3. _____
   A. Positive-pressure type
   B. Injector-type

4. Complete the following information. Refer to Figure 12-13 for a positive-pressure torch and to Figure 12-16 for an injector-type torch.

   drill size tip orifice: # _____ (positive-pressure)   # _____ (injector-type)

   psig oxygen pressure: _____ (positive-pressure)   _____ (injector-type)

   psig acetylene pressure: _____ (positive-pressure)   _____ (injector-type)

5. Place the metal between two firebricks, so only the outer edges are supported by the bricks.

6. Turn on the outfit, light the flame, and adjust the flame to neutral.

7. If you are right-handed, start the weld pool at the right end of the metal. If you are left-handed, start at the left end of the metal.

8. Point the flame in the direction of travel. Keep the flame tip at the correct height above the base metal. An angle of 35°–45° should be maintained between the torch tip and the base metal. This angle may be varied up or down to heat or cool the weld pool if it is too narrow or too wide.

9. Complete one weld pool. Using your pliers to pick the metal up, inspect the bottom side for good, even penetration. The weld pool should be straight and have a uniform width.

10. Finish the other four weld pools. Use the lines drawn on the metal to keep the weld pool going in a straight line.

11. Inspect each weld pool as it is made, and attempt to correct any defect in your next weld pool.

12. If necessary, try again with another piece of metal until five weld pools of good quality are complete on one piece.

Instructor's initials: _____

Job 12B-2

Square-Groove Weld on an Outside Corner

Name _____ Date _____

Class _____ Instructor _____

> **Learning Objective**
> - In this job, you will make a square-groove weld on an outside corner without the use of a welding rod.

1. After completing Job 12B-1, obtain two pieces of mild steel that measure 1/16" × 1 1/2" × 5" (1.6mm × 40mm × 125mm).

2. Arrange the pieces as shown in the following image to form an outside corner joint. The vertical piece should extend beyond the horizontal piece by 1/16" (1.6mm).

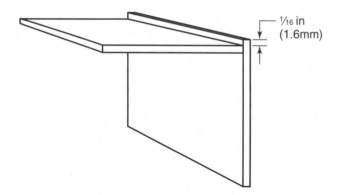

3. Tack weld the parts in three places, once at each end and once in the middle. This is done by melting down the overlap until it melts into the horizontal piece.

4. Start your weld from the end closest to your welding hand. Apply the torch to the overlapped edge and melt it down into the horizontal piece. A crescent-shaped or zigzag motion is suggested. Be certain to create a weld pool in the horizontal piece to permit the overlapped edge to melt (fuse) into it.

5. Make three of these outside corner joints without a welding rod.

> **Inspection**
> The weld bead should be slightly convex, with a uniform width, and with a small amount of penetration showing on the inside of the corner.
>
> Instructor's initials: _____

206   Modern Welding Lab Workbook   Copyright by The Goodheart-Willcox Co., Inc.

Lesson 12C

Oxyfuel Gas Welding—Welding Mild Steel in the Flat Welding Position

Name _____ Date _____

Class _____ Instructor _____

Learning Objective

● You will be able to select the correct welding rod. You will be able to select the correct tip size and welding gas pressure for a given metal thickness. You will also be able to weld several joints in the flat welding position.

Instructions

Carefully read Headings 12.3.4 through 12.3.6, 12.4, and 12.7. Also study Figures 12-27 through 12-46 and 12-59 through 12-61 in the text. Then answer the following questions.

1. A welding rod should be held within _____" (_____mm) of the flame tip and 1/16" to 1/8" (1.6mm to 3.2mm) of the surface of the weld pool to keep it preheated.

    1. _____

2. The shape of a normal weld bead should be _____.
    A. concave
    B. convex
    C. flat

    2. _____

3. Welding rod is added to the weld pool by _____.
    A. allowing it to drip from the end of the molten rod
    B. touching it to the base metal ahead of the weld pool
    C. dipping it into the forward edge of the molten weld pool
    D. touching it to the base metal just behind the weld pool
    E. melting it off the end with the torch flame

    3. _____

4. List three reasons why a rod smaller than recommended should *not* be used.

    _____
    _____
    _____

5. Welding rod is only added to the lap and inside corner joint when the weld pool runs forward at the outer edges to form a(n) ____-shaped weld pool.
   A. T
   B. C
   C. Z
   D. O
   E. S

5. _____

6. Define a weld made in the flat welding position.

   _____
   _____
   _____

7. List three methods used to maintain the correct position of welded workpieces as they cool and solidify.

   _____
   _____
   _____

8. Complete the drawing of a V-groove butt weld in progress. Show the "keyhole" clearly in your work.

9. A work angle of ____ should be used for welding a butt joint in the flat welding position.
   A. 0°
   B. 15°
   C. 30°
   D. 45°

9. _____

10. When a butt joint is welded in the flat position, the welding rod should be held at an angle of ____ from the surface of the base metal.
    A. 0°
    B. 30-60°
    C. 15-45°
    D. 60-75°

10. _____

Job 12C-1

Fillet Weld on a Lap Joint in the Flat Welding Position

Name _____ Date _____

Class _____ Instructor _____

> **Learning Objective**
> - In this job, you will produce a fillet weld on a lap joint in the flat welding position.

1. Obtain six pieces of mild steel that measure 3/32″ × 1 1/2″ × 5″ (2.4mm × 40mm × 125mm).

2. For this metal, use the following:

    # _____ drill tip size

    _____″ diameter welding rod

    _____ psig oxygen pressure

    _____ psig acetylene pressure

3. Tack weld the joint in three places on both sides.

4. Make one of the following fillet welds.

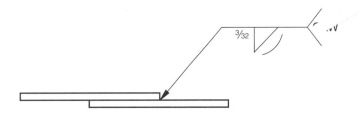

> **Inspection**
> 1. Your weld bead should be even in width and slightly convex, with smooth and evenly spaced ripples in the beads.
> 2. After inspecting one sample, make two more lap welds on the remaining pieces.
> 3. Use your final weldment for a test sample. Place the weldment into a vise. The weld should be just above the top of the vise jaws.
>   - **Caution:** The vise should have a safety screen around it to prevent the metal from flying across the room when it fails.
> 4. Bend one piece away from the other using a hammer. Continue to bend the piece back and forth until it fails. A strong weld will tear away from the bottom piece. The weld should come away from the bottom piece with some metal from the surface. If it does not, the bottom surface was not melted enough before the welding rod was added, and the weld fusion is poor.
>
>   Instructor's initials: _____

Job 12C-2

Fillet Weld on an Inside Corner in the Flat Welding Position

Name _____ Date _____

Class _____ Instructor _____

Learning Objective
- In this job, you will make a fillet weld on an inside corner joint in the flat welding position.

1. Obtain six pieces of mild steel that measure 1/8″ × 1 1/2″ × 5″ (3.2mm × 40mm × 125mm).

2. To weld this metal, use the following:
   - # _____ drill tip size
   - _____″ diameter welding rod
   - _____ psig oxygen pressure
   - _____ psig acetylene pressure

3. Tack weld the pieces three times on each side.

4. Make three of the following welds.

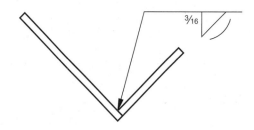

Inspection
All welds should be straight. The beads should be slightly convex and even in width, with smooth, evenly spaced ripples in the beads.

Instructor's initials: _____

210   Modern Welding Lab Workbook

Job 12C-3

Square-Groove Weld on an Outside Corner in the Flat Welding Position

Name _____ Date _____

Class _____ Instructor _____

> **Learning Objective**
>
> In this job, you will make a square-groove weld on an outside corner joint in the flat welding position.

1. Obtain six pieces of 3/32" (2.4mm) mild steel.

2. To weld this metal, use the following:
   - # _____ drill tip size
   - _____ " diameter welding rod
   - _____ psig oxygen pressure
   - _____ psig acetylene pressure

3. Tack weld about 3" (75mm) apart on the weld side only.

4. Make three of the following welds.

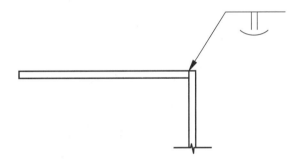

> **Inspection**
>
> All welds should be straight and slightly convex, with even width and smooth, even ripples in the bead. There should be little or no penetration showing on the other side of the weld.
>
> Instructor's initials: _____

Job 12C-4

Square-Groove Weld on a Butt Joint in the Flat Welding Position

Name _____ Date _____

Class _____ Instructor _____

> **Learning Objective**
> - In this job, you will make a square-groove weld on a butt joint in the flat welding position.

1. Obtain six pieces of mild steel that measure 1/16″ × 1 1/2″ × 5″ (1.6mm × 40mm × 125mm).

2. For this weld, use the following:

    # _____ drill tip size

    _____ ″ diameter welding rod

    _____ psig oxygen pressure

    _____ psig acetylene pressure

3. Tack weld about every 3″ (75mm) on the weld side.

4. Make three welds of the welds shown in the following image.

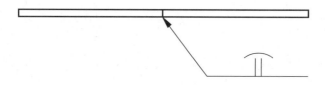

> **Inspection**
> All welds should be straight and slightly convex, with even width and smooth, even ripples in the bead. There should be little or no penetration showing on the other side of the weld.
>
> Use your last joint as a test sample. Place it into a vise so the middle of the weld bead is even with top of the vise jaws. Bend the metal back and forth until it fails. Always make the bend toward the screen or curtain. A weldment with a good weld will break in the base metal and not through the weld area.
>
> ■ **Caution:** Whenever a destructive test is made in a vise, the vise must have a screen or curtain around it.
>
> Instructor's initials: _____

Job 12C-5

Fillet Weld on an Inside Corner in the Flat Welding Position

Name _____ Date _____

Class _____ Instructor _____

> **Learning Objective**
>
> - In this job, you will make a fillet weld on an inside corner and a V-groove weld on an outside corner in the flat welding position.

1. Obtain four pieces of 1/4" × 1 1/2" × 5" (6.4mm × 40mm × 125mm) mild steel.

2. For this metal, use the following:

    # _____ drill tip size

    _____ " diameter welding rod

    _____ psig oxygen pressure

    _____ psig acetylene pressure

3. Bevel the edges as shown in the drawing.

    Caution: Before using the grinder, be certain you have passed the grinder safety test and are wearing grinding goggles.

4. Use 1/16" (1.6mm) steel as spacers while you tack weld the pieces. If possible, remove the spacers after tack welding is complete.

5. Perform the welds shown in the following image to complete the two weldments.

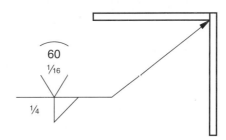

> **Inspection**
>
> The weld beads should be the correct size. The beads should be straight with uniform width and smooth, even ripples.
>
> Instructor's initials: _____

Job 12C-6

Fillet Weld on a Lap Joint in the Flat Welding Position

Name _____ Date _____

Class _____ Instructor _____

Learning Objective
- In this job, you will make a fillet weld on a lap joint in the flat welding position.

1. Obtain four pieces of 1/4" × 1 1/2" × 5" (6.4mm × 40mm × 125mm) mild steel.

2. For this metal, use the following:

   # _____ drill tip size

   _____ " diameter welding rod

   _____ psig oxygen pressure

   _____ psig acetylene pressure

3. Tack weld the joint in three places on each side.

4. Make two of the weldments shown in the following image.

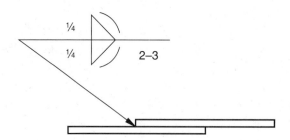

Inspection
Inspect visually. Each bend must be straight, with uniform width and smooth, even ripples.

Instructor's initials: _____

Job 12C-7

V-Groove Weld on a Butt Joint in the Flat Welding Position

Name _____ Date _____

Class _____ Instructor _____

> **Learning Objective**
> - In this job, you will make a V-groove weld and a bevel-groove weld on a butt joint in the flat welding position.

1. Obtain four pieces of mild steel that measure 1/4" × 1 1/2" × 5" (6.4mm × 40mm × 125mm).

2. Prepare the edges of the metal as needed.

   Caution: Before using the grinder, be certain you have passed the grinder safety test and are wearing grinding goggles.

3. Make the welds shown in the following drawing.

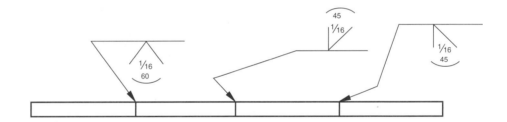

> **Inspection**
> Each butt weld must have full penetration throughout its length, an even bead width, a convex bead, and smooth, even ripples in the bead.
>
> Instructor's initials: _____

Lesson 12D

Oxyfuel Gas Welding—Welding Mild Steel in the Horizontal Welding Position

Name _____ Date _____

Class _____ Instructor _____

Learning Objective
- You will be able to weld the various types of joints in the horizontal welding position.

Instructions
Carefully read Headings 12.3.4 through 12.3.6, 12.4, 12.5, 12.5.1, and 12.7. Also study Figures 12-47 through 12-49 and 12-57 in the text. Then answer the following questions.

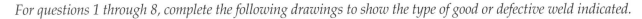

For questions 1 through 8, complete the following drawings to show the type of good or defective weld indicated.

1. A properly made weld with complete penetration and proper fusion at the toes of the weld.

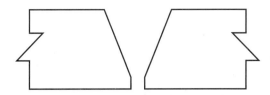

2. A weld with less than 100% penetration, with proper fusion, and overlapping at the toes of the weld.

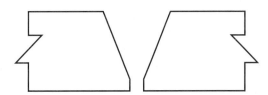

3. A properly made weld with good fusion at the toes of the weld and adequate penetration for a lap joint.

4. A weld with incomplete fusion at the toe of the weld.

5. A fillet weld on a T-joint. This weld has undercutting on the vertical piece and has a smaller-than-normal weld size. Show the ideal weld size as a hidden (dashed) line.

6. A butt weld, with undercutting at the toes and a weld face below the metal surface.

7. A fillet weld, with undercutting on the lower piece. Show a weld with a convex bead.

8. A weld with 100% penetration but with a concave bead.

9. Define the horizontal welding position. _____
_____

10. To prevent the molten metal in a weld from sagging while making a weld in the horizontal position, point the torch tip and welding flame _____.
    A. straight along the weld line
    B. slightly downward from the weld axis (line)
    C. perpendicular to the base metal
    D. slightly upward from the weld axis (line)
    E. at the lower piece of metal

10. _____

Job 12D-1

Fillet Weld on a Lap Joint in the Horizontal Welding Position

Name _____ Date _____

Class _____ Instructor _____

Learning Objective
- In this job, you will make a fillet weld on a lap joint in the horizontal welding position.

1. Obtain four pieces of mild steel that measure 3/32″ × 1 1/2″ × 5″ (2.4mm × 40mm × 125mm).

2. Use the following for these welds:

    # _____ drill tip size

    _____″ diameter welding rod

    _____ psig oxygen

    _____ psig acetylene

3. Tack weld the pieces to create the weldment shown. Make all tack welds in the flat welding position. Each joint should be tack welded at three places.

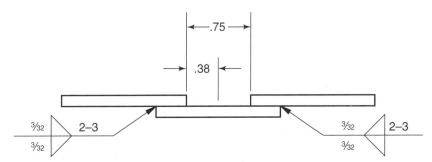

4. Place the metal in a positioning fixture or prop it against firebricks.

5. Make the welds shown in the drawing. Reposition the weldment as needed so all welds are made in the horizontal welding position.

Inspection
Visually check each fillet weld. The beads should be straight, convex, have evenly spaced ripples, and have good fusion between the bead and surface. Two of the welds are intermittent welds that should be 2″ long and spaced 3″ center-to-center.

Instructor's initials: _____

Copyright by The Goodheart-Willcox Co., Inc.   Modern Welding Lab Workbook

Job 12D-2

Fillet Weld on a T-Joint in the Horizontal Welding Position

Name _____ Date _____
Class _____ Instructor _____

Learning Objective

- In this job, you will make a fillet weld on a T-joint in the horizontal welding position.

1. Obtain three pieces of mild steel that measure as follows:
   Two – 1/16" × 1 1/2" × 5" (1.6mm × 40mm × 125mm)
   One – 1/16" × 3" × 5" (1.6mm × 75mm × 125mm)

2. Use the following for these welds:

   # _____ drill tip size

   _____ " diameter welding rod

   _____ psig oxygen pressure

   _____ psig acetylene pressure

3. Assemble two of the pieces to make the weldment shown in the following drawing. Tack weld three places on each side of the T-joint. This can be done with the metal in the flat position.

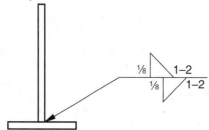

4. Place the weldment into the positioning fixture or prop it against firebricks.

5. Make the welds shown in the horizontal position. The weld axis must be horizontal and the weld face at 45°.

6. After completing Step 5, add the third piece of metal and tack weld it in place. Tack welds can be made in the flat welding position. Make the following welds to complete the weldment.

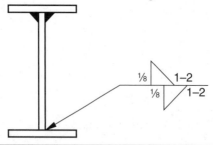

Inspection

Make a visual inspection. The dimensions of the staggered intermittent welds should be approximately as shown in the previous images.

Instructor's initials: _____

Job 12D-3

Square-Groove Weld on a Butt Joint in the Horizontal Welding Position

Name _____ Date _____

Class _____ Instructor _____

Learning Objective
- In this job, you will make a square-groove weld on a butt joint in the horizontal welding position.

1. Obtain four pieces of mild steel that measure as follows:
   Two 1/16″ × 1 1/2″ × 5″ (1.6mm × 40mm × 125mm)
   Two 1/8″ × 1 1/2″ × 5″ (3.2mm × 40mm × 125mm)

2. For the 1/16″ (1.6mm) thick steel (weld #1), use the following:
   # _____ drill tip size
   _____″ diameter welding rod
   _____ psig oxygen pressure
   _____ psig acetylene pressure

3. For the 1/8″ (3.2mm) thick steel (weld #2), use the following:
   # _____ drill tip size
   _____″ diameter welding rod
   _____ psig oxygen pressure
   _____ psig acetylene pressure

4. Arrange the 1/16″ thick pieces as shown in the following drawing. Tack weld the joint in three places. The tack welds can be made in the flat welding position.

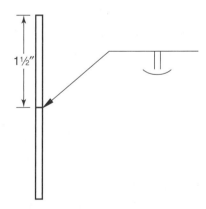

5. Place the weldment into the positioning fixture.

6. Complete the weldment as shown in the previous drawing. The square-groove weld must be completed in the horizontal welding position.

7. Next, assemble the 1/8″ thick pieces as shown in the following drawing. Be sure to set the desired gap opening, and then tack weld the square-groove joint in three places. Tack welds can be completed in the flat welding position.

8. Place the weldment in a positioning fixture so the weld can be made in the horizontal welding position.

9. Complete the weld as indicated in the drawing.

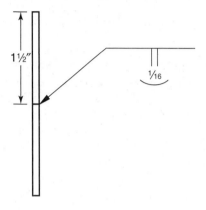

Inspection

Make a visual inspection. Be certain that the butt welds have 100% penetration. Bend each weld until it fails. In a properly made butt weld, the failure must occur in the base metal and not in the weld metal.

■ **Caution:** Wear safety glasses while performing the bend test. Also, bend the welds only toward the curtain or screen around the vise.

Instructor's initials: _____

Lesson 12E

Oxyfuel Gas Welding—Welding Mild Steel in the Vertical Welding Position

Name _____ Date _____

Class _____ Instructor _____

Learning Objective
- You will be able to describe methods of welding thick metal sections. You will also be able to make welds in the vertical welding position.

Instructions
Carefully read Headings 12.3.6, 12.5.2 and 12.6 of the text. Also study Figures 12-29 through 12-31, 12-50, 12-51, and 12-55 through 12-58 in the text. Then answer the following questions.

1. The _____ keeps the molten metal in the vertical weld from sagging and falling out of the weld pool.
   A. force of gravity
   B. force of the welding gas coming from the tip
   C. fact that the metal cools rapidly
   D. action of the hot welding rod
   E. air circulating around the weld

   1. _____

2. For welding in the vertical position, the torch should be pointed upward with a travel angle of approximately _____.
   A. 15°–30°
   B. 30°–45°
   C. 45°–60°
   D. 50°–65°
   E. 60°–75°

   2. _____

3. Name the various areas and measurements of the V-groove weld below.
   A. _____
   B. _____
   C. _____
   D. _____
   E. _____

4. Name the nine butt joint welds shown.

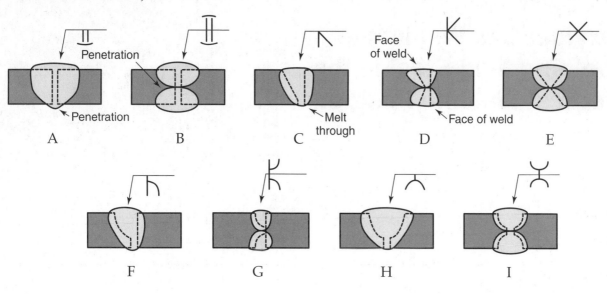

A. _____
B. _____
C. _____
D. _____
E. _____
F. _____
G. _____
H. _____
I. _____

5. Directing the flame in a backhand manner tends to _____ the completed weld, which relieves the welding stresses.

5. _____

6. Which of the following statements is true for multiple-pass or step-pass welds?
    A. They are generally made on thick metal.
    B. They permit beads of a size that are easily handled.
    C. They permit welds thin enough to allow impurities to escape before the weld solidifies.
    D. All of the above.

6. _____

7. Backhand welding requires that the torch be held at a(n) _____ angle of 30°–45°.

7. _____

8. Describe "backhand welding."

_____
_____
_____
_____

224   Modern Welding Lab Workbook

Lesson 12E Oxyfuel Gas Welding—Welding Mild Steel in the Vertical Welding Position

Name _____

9. Define a "multiple-pass weld."

   _____
   _____
   _____
   _____

10. Define the term "vertical welding position."

    _____
    _____
    _____
    _____

Job 12E-1

Fillet Weld on a Lap Joint in the Vertical Welding Position

Name _____ Date _____

Class _____ Instructor _____

Learning Objective
- In this job, you will make a fillet weld on a lap joint in the vertical welding position.

1. Obtain three pieces of mild steel that measure 1/16″ × 1 1/2″ × 5″ (1.6mm × 40mm × 125mm).

2. Use the following for these welds:
   # _____ drill tip size
   _____ ″ diameter welding rod
   _____ psig oxygen
   _____ psig acetylene

3. Assemble the pieces as shown in the following drawing. Tack weld each of the joints at three places. This can be done in the flat welding position.

4. Place the weldment into a positioning fixture so the weld axes are vertical. You can prop the weldment against firebricks if a positioning fixture is not available.

5. Make the four welds shown in the drawing. Reposition the weldment as necessary to make all the welds in the vertical welding position and from the bottom up.

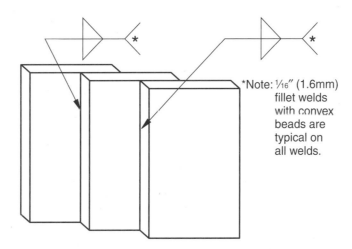

*Note: 1/16″ (1.6mm) fillet welds with convex beads are typical on all welds.

Inspection
Each fillet weld should be straight, convex, and have evenly spaced ripples. There should be good fusion between the bead and the surface.

Instructor's initials: _____

Job 12E-2

Fillet Weld on a T-Joint in the Vertical Welding Position

Name _____ Date _____

Class _____ Instructor _____

> **Learning Objective**
> - In this job, you will weld a fillet weld on a T-joint in the vertical welding position.

1. Obtain three pieces of mild steel that measure 1/8″ × 1 1/2″ × 5″ (3.2mm × 40mm × 125mm).

2. Use the following for these welds:

    # _____ drill tip size

    _____″ diameter welding rod

    _____ psig oxygen

    _____ psig acetylene

3. Assemble and tack weld the weldment as shown in the following drawing. Tack welding can be done in the flat welding position.

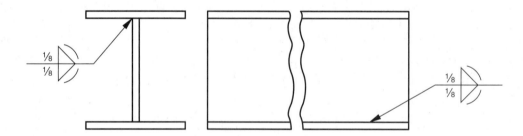

4. Place the weldment in a welding fixture or prop it against several firebricks. The weld axes must be vertical.

5. Make all four welds shown in the drawing in the vertical welding position.

> **Inspection**
> All welds should be 1/8″ (3.2mm) in size, convex, straight, with evenly spaced ripples, and with good fusion at the toes of the weld.
>
>   Instructor's initials: _____

Job 12E-3

V-Groove Weld on an Outside Corner in the Vertical Welding Position

Name _____ Date _____

Class _____ Instructor _____

> Learning Objective
> • In this job, you will make a V-groove weld on an outside corner joint in the vertical welding position for this job.

1. Obtain four pieces of mild steel that measure 1/8″ × 1 1/2″ × 5″ (3.2mm × 40mm × 125mm).

2. Use the following for these welds:

    # _____ drill tip size

    _____ ″ diameter welding rod

    _____ psig oxygen

    _____ psig acetylene

3. Assemble and tack weld the weldment shown in the following drawing. Tack welding can be done in the flat welding position.

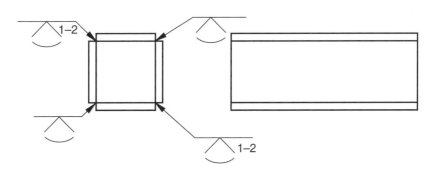

4. Place the weldment in a welding fixture or prop it against several firebricks. The weld axis must be vertical.

5. Make all four welds in the vertical welding position.

> **Inspection**
> Each weld bead should be convex and straight with evenly spaced ripples. The weld should fill the V-groove. Each V-groove weld must have 100% penetration. Two welds should have intermittent beads, 1″ long and spaced 2″ on center.
>
> Instructor's initials: _____

Lesson 12F

Oxyfuel Gas Welding—Welding Mild Steel in the Overhead Welding Position

Name _____   Date _____
Class _____   Instructor _____

Learning Objective
- You will be able to practice the safety precautions required for overhead welding. You will also be able to weld in the overhead welding position.

Instructions
Carefully read Headings 12.5.3 of the text. Also study Figures 12-52 through 12-54 in the text. Then answer the following questions.

1. Define overhead welding.
   _____
   _____
   _____
   _____

2. To prevent hot metal from being caught, pant legs should not have _____ on them.   2. _____

3. *True or False?* Overheating will cause the metal to become too fluid and may cause the molten metal to fall.   3. _____

4. The momentary movement of the flame away from the weld pool and back again is called a(n) _____ motion.   4. _____

5. The motion in Question 4 is used to _____.   5. _____
   A. superheat the weld pool
   B. cool the weld pool
   C. allow space for the welding rod
   D. allow time to insert the welding rod into the welding pool
   E. create a larger weld pool

6. If the torch tip becomes plugged by molten metal, it should be cleaned using a(n) _____.   6. _____

Modern Welding Lab Workbook   231

7. What is *surface tension*?

___

8. What two forces act to keep the molten metal from falling out of the weld pool?

___

9. List three pieces of protective clothing that must be worn to protect the head, hands, and body during overhead welding.

___

10. What safety equipment should be worn during overhead welding to protect a welder's legs and feet from falling molten metal?

___

Job 12F-1

Square-Groove Weld on a Butt Joint, Fillet Weld on a Lap Joint, and Fillet Weld on a T-Joint in the Overhead Welding Position

Name _____ Date _____

Class _____ Instructor _____

> **Learning Objective**
> - In this job, you will make square-groove welds on a butt joint, fillet welds on a lap joint, and fillet welds on a T-joint. All welds will be made in the overhead position.

1. Obtain five pieces of mild steel that measure 1/16" × 1 1/2" × 5" (1.6mm × 40mm × 125mm).

2. Use the following for these welds:

    # _____ drill tip size

    _____" diameter welding rod

    _____ psig oxygen

    _____ psig acetylene

3. Assemble and tack weld the weldment shown in the following drawing. This can be done in the flat welding position.

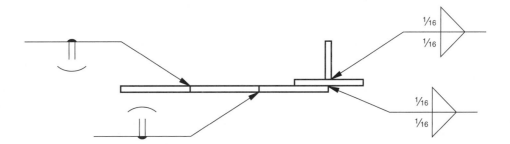

4. Place the weldment into a positioning fixture. Adjust the positioning fixture so that the joints are high enough to permit overhead welding.

5. Make all the welds indicated in the drawing. Reposition the weldment to ensure that each weld is made in the overhead welding position.

> **Inspection**
> Check each weld. The butt joints should have 100% penetration. All welds should be convex, with evenly spaced ripples, and with good fusion at the toes of the weld.
>
> Instructor's initials: _____

Instructor's initials: _____
Butt joint grade: _____

Instructor's initials: _____
Lap joint grade: _____

Instructor's initials: _____
T-joint grade: _____

Lesson 13

Oxyfuel Gas Cutting Equipment and Supplies

Name _____ Date _____
Class _____ Instructor _____

Learning Objective
• You will be able to identify and correctly use the equipment and supplies used in oxyfuel gas cutting.

Instructions
Carefully read Headings 13.1 through 13.6 of the text. Also study Figures 13-1 through 13-22 in the text. Then answer the following questions.

1. Name the parts of the oxyacetylene cutting outfit shown in the following image.

   A. _____
   B. _____
   C. _____
   D. _____
   E. _____
   F. _____

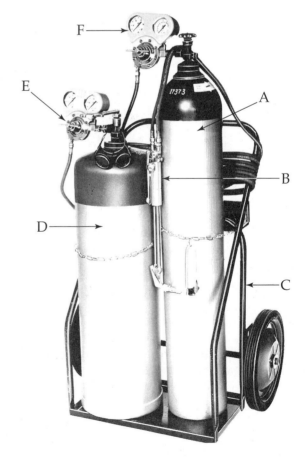

Copyright by The Goodheart-Willcox Co., Inc.   Modern Welding Lab Workbook   235

2. Name the parts indicated on the following image of a standard oxyfuel gas cutting torch.

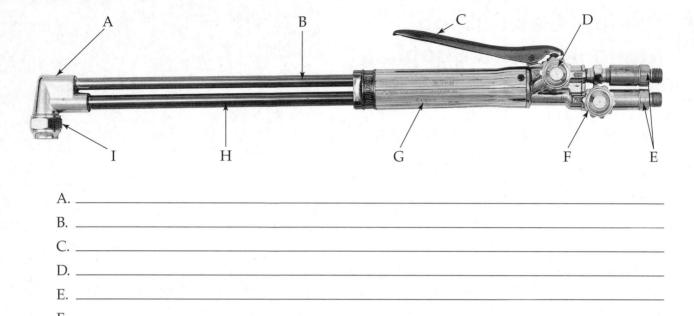

A. _____
B. _____
C. _____
D. _____
E. _____
F. _____
G. _____
H. _____
I. _____

3. Name the parts indicated on the following image of a cutting torch attachment.

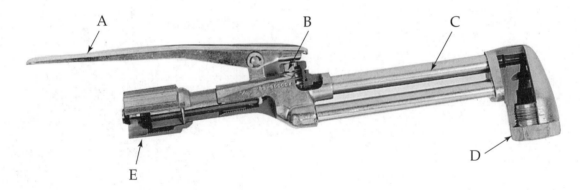

A. _____
B. _____
C. _____
D. _____
E. _____

4. Cutting tips have one oxygen cutting orifice and one or more _____ orifices.

4. _____

Lesson 13 Oxyfuel Gas Cutting Equipment and Supplies

Name _____

5. Describe the application of the two cutting torch tips shown.

   A. _____

   B. _____

6. What is the name of the tip tool shown in the following image?

   _____

7. The welder in the following image is using a(n) _____ to help cut a circle with the torch.

   _____

8. A heavy duty oxygen regulator is the best choice for oxyacetylene cutting because it _____.
   A. has two stages and heavy duty springs
   B. has a high capacity orifice
   C. permits a higher gas flow
   D. is capable of handling a working pressure of 100–150 psig
   E. All of the above.

8. _____

9. Mechanical guides can be used to control the _____ of the torch.

9. _____

10. Which of the following is *not* true of an electric motor-driven carriage?
    A. The track may be flexible to permit a curve to be followed.
    B. The torch is mounted off the side of the carriage.
    C. The track is attached to the weldment with magnets.
    D. A carriage typically has two wheels.

10. _____

Lesson 14

Oxyfuel Gas Cutting—Cutting Steel

Name _____ Date _____
Class _____ Instructor _____

Learning Objective
- You will be able to cut and gouge various thicknesses of steel.

Instructions
Carefully read Headings 13.5.2 and 14.1 through 14.10 of the text. Also study Figures 14-1 through 14-39 in the text. Then answer the following questions.

1. It is possible for an oxyacetylene gouging tip to gouge, but not cut through, metal because a _____ is used.
   A. lower oxygen cutting pressure
   B. smaller cutting oxygen orifice
   C. higher flame temperature
   D. All of the above.

   1. _____

2. The procedure for purging an oxyacetylene cutting outfit and lighting a manual positive-pressure cutting torch follows. The nine steps are not in the correct order. Place them in the correct order by writing the correct letter in the appropriate blank.
   A. Open the acetylene torch valve one turn. Slowly turn in the acetylene regulator adjusting screw until the low-pressure acetylene gauge indicates working pressure corresponding to tip size. Close the torch acetylene valve.
   B. Open the oxygen cylinder valve very slowly until the regulator high-pressure gauge reaches its maximum reading. Then, turn the cylinder valve all the way open to close the double seating (backseating) valve. Open the acetylene cylinder valve slowly 1/4 to 1/2 turn. Leave the acetylene cylinder valve wrench in place.
   C. Open the torch acetylene valve until the acetylene flame jumps away from the end of the tip slightly and back again when the torch is given a shake or whipping action.

   1. _____
   2. _____
   3. _____
   4. _____
   5. _____
   6. _____
   7. _____
   8. _____
   9. _____

(Continued)

Copyright by The Goodheart-Willcox Co., Inc. Modern Welding Lab Workbook 239

D. Open the torch acetylene valve approximately 1/16 to 1/8 turn. Then, use a flint lighter to ignite the acetylene.
E. Check the condition of all equipment. Inspect the outfit to ensure that it has been assembled correctly.
F. Turn the regulator adjusting screws all the way out (regulator closed).
G. Open the torch oxygen valve one full turn. Next, press the cutting oxygen lever and adjust the oxygen regulator to give the desired oxygen working pressure. Close the torch oxygen valve and release the cutting oxygen lever.
H. After setting the working pressures, check the low-pressure gauge readings.
I. Open the torch oxygen valve and adjust it to obtain neutral preheating flames. Press the cutting oxygen lever and readjust the preheating flames if necessary.

3. *True or False?* After a cut on cast iron is completed, the casting is cooled rapidly if gray cast iron is desired.

3. _____

4. The approximate oxygen and acetylene pressures to use when cutting 1" (25.4mm) mild steel with a positive-pressure torch are:

   Oxygen psi _____

   Acetylene psi _____

5. A bell-mouthed kerf is generally caused by ____.
   A. too much acetylene pressure
   B. an oxidizing flame
   C. too much oxygen pressure
   D. a carburizing flame
   E. too fast a torch motion

5. _____

6. List seven variables that a welder must monitor and adjust when cutting metals with refractory oxides.
   _____
   _____
   _____
   _____
   _____
   _____
   _____

7. List three ways to correct excessive drag during oxyfuel gas cutting.
   _____
   _____
   _____

8. *True or False?* Oxyfuel gas cutting operations result in considerable sparking.

8. _____

Lesson 14 Oxyfuel Gas Cutting—Cutting Steel

Name _____

9. What actions should be taken to displace gases in a tank that is undergoing a cutting operation?
_____
_____
_____
_____
_____

10. In a gouging tip, there are ____ or ____ preheat orifices to provide an even distribution of the preheat flames.

10. _____
    _____

Job 14-1

Cutting Mild Steel with an Oxyfuel Gas Torch

Name _____ Date _____
Class _____ Instructor _____

> **Learning Objective**
> - In this job, you will cut steel using an oxyfuel gas cutting torch.

1. Obtain a piece of mild steel that measures 1/4″ × 8″ × 8″ (6.4mm × 200mm × 200mm).
2. Obtain a second piece of mild steel that is thicker than 1/4″ (6.4mm), preferably 1″ (25mm) thick. The thick piece should be at least 3″ (75mm) wide and approximately 8″ (200mm) long.
   - **Note:** The width, length, and thickness can vary according to the dimensions of steel that is available.
3. Using a soapstone, make seven lines to mark off eight 1″ (25mm) wide strips on the 1/4″ (6.4mm) thick plate.
4. Place the 1/4″ (6.4mm) plate on the cutting table. Place the plate on firebricks so the cutting table is not damaged as each cut is made. The area to be cut should hang over the edge of the firebrick.
5. To make cuts with a positive-pressure torch, use the following variables:
   _____ cutting oxygen orifice size
   _____ psig oxygen pressure
   _____ psig acetylene pressure
   _____ in/min cutting speed
6. Cut four pieces 1″ (25mm) wide with square edges. Make cuts on the remaining lines so that each of the remaining four pieces each have a 45° bevel or chamfer on the edges.
7. Place the thicker plate on the cutting table. Arrange the plate so the table is not damaged.
8. The plate thickness used is ____″ (____mm) thick.   8. _____
9. To perform a cut on the thicker metal, use the following variables:
   _____ cutting oxygen orifice size
   _____ psig oxygen pressure
   _____ psig acetylene pressure
   _____ in/min cutting speed
10. Using a soapstone, make two lines to mark off three pieces 1″ (25mm) wide.
11. Make the first cut with a square edge. Make the second cut using a 45° bevel or chamfered edge.

> **Inspection**
> Each cut should be straight with a clean, smooth, square edge. The drag lines should be nearly vertical and not wavy, rough, or irregular. There should be no slag adhering to the edge.
>
> Instructor's initials: _____

Job 14-2

Cutting Shapes with an Oxyfuel Gas Torch

Name _____ Date _____

Class _____ Instructor _____

Learning Objective
- In this job, you will cut shapes using a manual oxyfuel gas torch.

1. Obtain two pieces of plain carbon steel that measure 3/8" × 2" × 6" (9.6mm × 50mm × 150mm).
2. Using a soapstone, draw the shape shown on the following drawing on both pieces of metal.

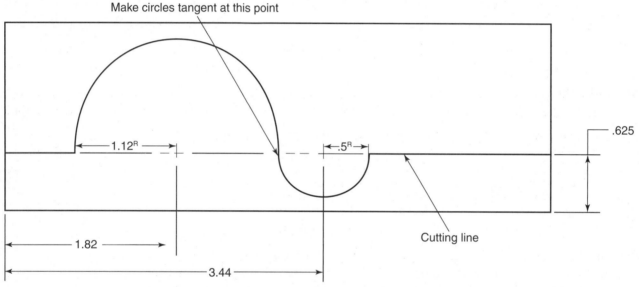

3. Place one piece of metal on the cutting table. Prepare the cutting outfit to cut the shape drawn on the base metal.
4. To make the cuts with a positive-pressure cutting torch, use the following variables:

    _____ cutting oxygen orifice size

    _____ psig oxygen pressure

    _____ psig acetylene pressure

    _____ in/min cutting speed

5. Cut the shape drawn on the base metal.
6. Cut the second piece of metal in the same manner.

Inspection
The continuous cut should have a clean, smooth, square edge from beginning to end. The drag lines should be nearly vertical and not rough or too pronounced, showing a proper travel speed. If the cut is stopped and restarted, there should not be gouges at the restart point.

Instructor's initials: _____

Job 14-3

Removing Weld Reinforcement from the Face of a Weld

Name _____ Date _____

Class _____ Instructor _____

> **Learning Objective**
> - In this job, you will remove the weld reinforcement from the face of a weld using the oxyfuel gas gouging process.

1. Obtain three previously made welds with convex weld beads from the scrap metal container. Ask your instructor to approve your choices.

2. With the help of your instructor, select an appropriate gouging tip to accomplish this job. Install the gouging tip on the cutting torch of your cutting outfit.

3. To perform this gouging operation, set the following variables on your cutting station based on the thickness of the scrap metal you selected in Step 1.

   _____ preheating orifice size
   _____ cutting oxygen orifice size
   _____ psig (_____kPa) oxygen pressure
   _____ psig acetylene pressure
   _____ in/min cutting speed

4. Remove the weld reinforcement from one of the three welds.

5. Check the appearance of the base metal surface to determine if any changes are required to the oxygen and acetylene pressure, your technique, or travel speed.

6. Remove the weld reinforcement from the other two welds.

> **Inspection**
> The convex weld reinforcement should be completely removed as a result of the gouging operation. The base metal should be clean and smooth from the beginning of the gouging process to the end. The base metal can be flush or slightly convex where the weld bead or weld reinforcement was previously located.
>
> Instructor's initials: _____

Job 14-4

Setting Up and Inspecting a Semiautomatic Oxyfuel Gas Cutting Torch

Name _____ Date _____

Class _____ Instructor _____

> **Learning Objective**
> - In this job, you will set up and inspect a semiautomatic oxyfuel gas cutting torch mounted on a motorized carriage.

1. Check that the track used with your OFC electrically driven cutting torch carriage is laying flat on the cutting table.

2. Check that the wheels of the cutting torch carriage are in the tracks.

3. Place the base metal to be cut next to the carriage on the cutting table. Make sure the base metal is laying flat.

4. Mark the line to be cut.

5. Adjust the angle of the cutting torch by loosening the clamp that holds the torch on the carriage. A square cut or a chamfer of any angle can be made. After the torch-to-base metal angle is correct, tighten the torch.

6. Adjust the cutting torch in or out to align the torch tip with the cutting line.

7. Adjust the height of the cutting tip above the base metal.

8. After these adjustments are made, run the carriage along the track to ensure that the torch and hoses are not hitting anything as the cut progresses along the cutting line.

9. Using the proper size open-end wrench, check that all hose fittings are tight on the carriage, torch, and regulators.

    ■ **Caution:** Do not overtighten the brass fittings.

10. Check the hoses carefully for worn or cut areas.

11. Turn on the cutting outfit and set the oxygen and acetylene working pressures on the regulators.

12. Set the forward travel speed on the torch carriage and engage the clutch to test the travel speed.

13. The cutting torch carriage and torch are now ready to make the cut.

14. Have the instructor check your inspection and machine set-up.

> **Inspection**
> Inspect the work of the student. Ensure that the student has carefully followed all of the above steps prior to beginning to make cuts with the carriage-mounted OFC torch.
>
>   Instructor's initials: _____

Job 14-5

Making Straight-Line Cuts on Steel to Produce Square and Beveled Edges

Name _____ Date _____

Class _____ Instructor _____

> **Learning Objective**
>
> ● In this job, you will make straight line cuts on plain carbon steel to produce square and beveled edges using an OFC torch mounted on a motorized carriage.

1. Obtain two pieces of plain carbon steel that measure 1/2″ × 8″ × 20″ (12.7mm × 200mm × 500mm).
2. Using a soapstone, mark four 2″ (50mm) wide strips on each piece.
3. To make the cuts with a positive-pressure cutting torch, use the following variables:

   _____ preheating orifice size
   _____ cutting oxygen orifice size
   _____ psig oxygen pressure
   _____ psig acetylene pressure
   _____ in/min cutting speed

4. Place the base metal on a cutting table. Make sure the base metal is laying flat.
5. Set up the electric cutting torch carriage using the settings listed in Step 3. Test the travel of the carriage.
6. Set the torch angle to cut a square edge in the kerf.
7. Light the cutting torch in the same manner used to light an OFW torch.
8. Engage the carriage clutch to begin the cut.
9. Cut the first piece on the three marked places.
10. Watch each cut as it progresses. Stop the carriage by disengaging the carriage clutch if the variables need to be changed to make a better cut.
11. Place the second piece of base metal on the cutting table and align the carriage and cutting torch.
12. Align the carriage and torch with the first line to be cut.
13. Set the torch so the cutting tip will cut a kerf at a 45° angle.
14. Make the first bevel-edged cut.
15. Examine the first bevel cut. Change any of the cutting variables if necessary to create an acceptable cut surface on the next two cuts.
16. Cut the next two bevels on the cutting lines marked.

> **Inspection**
>
> The continuous cut made with the OFC torch mounted on a motorized carriage should have a very clean and smooth square edge from beginning to end. The correct travel speed is proven if the drag lines are nearly vertical and not wavy, rough, or irregular. There should not be any slag adhering to the edge.
>
> Instructor's initials: _____

Lesson 15

Soldering

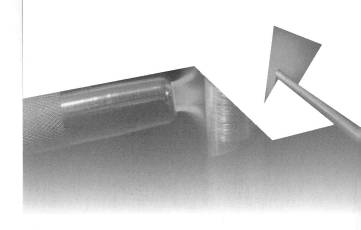

Name _____ Date _____
Class _____ Instructor _____

> **Learning Objective**
> • You will be able to describe the principles of soldering. You will also be able to perform various methods of soldering.
>
> **Instructions**
> *Carefully read Headings 15.1 through 15.8 of the text. Also study Figures 15-1 through 15-27 in the text. Then answer the following questions.*

1. The principle of _____ action is used to draw solder into the small clearance space between the pipe and fitting in a pipe joint.

   1. _____

2. What are the eutectic alloy percentage and the liquidus temperature for a tin-lead solder?
   A. _____ % lead
   B. _____ % tin
   C. _____ °F (_____ °C) liquidus temperature

   2. A. _____
      B. _____
      C. _____

3. What type of solder is used for soldering stainless steel food handling tables and equipment?

   3. _____

4. According to the AWS ANSI Z49.1 booklet, which elements contained in soldering alloys and fluxes have very low exposure limits?

   _____
   _____
   _____
   _____

5. *True or False?* A good soldering flux protects metal surfaces from oxidizing during the soldering process.

   5. _____

6. What does the term *wetting* mean? _____
   _____

Copyright by The Goodheart-Willcox Co., Inc.   Modern Welding Lab Workbook   249

7. A(n) _____ flux has the best cleaning action. 7. _____
   A. organic
   B. nonactive rosin
   C. rosin
   D. inorganic

8. What type of flux is corrosive during the soldering operation and noncorrosive after soldering? 8. _____

9. _____ rosin flux has the best cleaning action among rosin-based fluxes. 9. _____
   A. Nonactive
   B. Fully active
   C. Mildly active

10. Is a corrosive or noncorrosive solder recommended for soldering gold? 10. _____

11. The following general steps for torch soldering are out of order. Place the letters of the steps in the correct order in the spaces to the right.
    A. Remove any flux residue from the joint.
    B. Heat the metals to be soldered together.
    C. Apply the correct flux to the joint surfaces.
    D. Apply the proper amount of solder to the joint.
    E. Clean the surfaces to be soldered.

    1. _____
    2. _____
    3. _____
    4. _____
    5. _____

12. Name the following types of soldered joints.
    A. _____
    B. _____
    C. _____
    D. _____
    E. _____
    F. _____

13. Name three items that must be worn while working with cleaning solvents, acids, and pickling solutions.

    _____
    _____
    _____

Name _____

14. The process of adhering a very thin layer or film of solder to a metal surface prior to soldering is called _____.

14. _____

15. Describe the procedure for removing corrosive inorganic flux residues.

_____

_____

_____

_____

16. What soldering method is shown in the following drawing? _____

_____

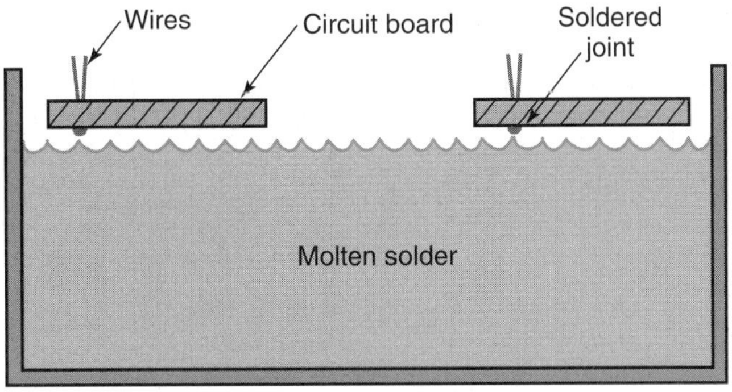

17. Which soldering method(s) generally use preformed and preplaced solder forms in the assembly to be soldered?
    A. Wave soldering
    B. Oven or infrared soldering
    C. Torch soldering
    D. Resistance and induction soldering
    E. Dip soldering

17. _____

18. _____ flux is recommended for stainless steel soldering.
    A. Water-soluble organic
    B. Fully active rosin
    C. Highly active inorganic
    D. Mildly active rosin
    E. Nonactive rosin

18. _____

19. List the three soldering alloys generally recommended for aluminum.

_____

_____

_____

20. *True or False?* In torch soldering, solder should be added to a cool joint by melting the solder with the torch.

20. _____

Job 15-1

Soldering Copper Fittings Used in Plumbing

Name _____ Date _____

Class _____ Instructor _____

> **Learning Objective**
>
> - In this job, you will solder copper fittings for plumbing drain lines using a tin-lead solder.
>
>   **Note:** Tin-lead solder must not be used on plumbing that will carry drinking water.

1. Obtain the following copper fittings:
   1/2" (12.7mm) cap
   1/2" (12.7mm) elbow
   1/2" (12.7mm) union (female half only)

   Note: If the fittings have already been assembled, unsolder them.

2. Obtain two lengths of 1/2" (12.7mm) copper tubing approximately 7" (180mm) long.

3. Use a corrosive or noncorrosive flux and a common 50/50 tin-lead solder.

4. Using an air-acetylene or propane torch and the correct soldering procedure, assemble the pipe and fittings as shown.

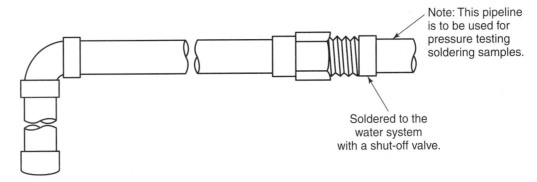

Note: This pipeline is to be used for pressure testing soldering samples.

Soldered to the water system with a shut-off valve.

> **Inspection**
>
> Assemble your soldered assembly to the water line with the union. Open the water valve and test for leaks. Turn off the water valve and remove your assembly.
>
> Instructor's initials: _____

Copyright by The Goodheart-Willcox Co., Inc.

Modern Welding Lab Workbook   253

Job 15-2

Soldering a Folded Seam

Name _____   Date _____

Class _____   Instructor _____

Learning Objective
- In this job, you will solder a folded metal seam.

1. Obtain two pieces of 28 gauge galvanized steel that measure 1 1/2" × 6" (40mm × 150mm).

2. Form seams on each piece to make the following assembly.

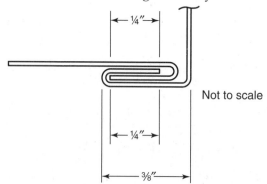

Not to scale

3. Clean the pieces.

4. Use a corrosive or noncorrosive flux and a common 50/50 tin-lead solder.

5. Using the correct soldering procedure, assemble and solder the joint.

Inspection
The proper amount of solder should be present. A thin film of solder should be visible in all folded seams.

Instructor's initials: _____

Lesson 16A

Brazing and Braze Welding Principles

Name _____ Date _____

Class _____ Instructor _____

Learning Objective
- You will be able to describe the principles of brazing and braze welding. You will also be able to choose the correct flux and brazing filler metal for a given job.

Instructions
Carefully read the introduction to Chapter 16 and Headings 16.1 through 16.5.1 of the text. Also study Figures 16-1 through 16-22 in the text. Then answer the following questions.

1. Which of the following statements is *true* of brazing? 1. _____
   A. It is done at a temperature below 840°F (450°C).
   B. Very thick layers of filler metal are used.
   C. The filler metal is distributed by capillary action.
   D. It is generally done on thick metal sections.

2. List three advantages of performing brazing or braze welding rather than welding.

   _____

   _____

   _____

3. The most important thing to do before applying flux prior to 3. _____
   braze welding or brazing is to _____ the metal surfaces.

4. List six of the ingredients typically found in a brazing flux.

   _____

   _____

   _____

Copyright by The Goodheart-Willcox Co., Inc.   Modern Welding Lab Workbook   255

5. Name the six braze welded joints shown.

   A. _____
   B. _____
   C. _____
   D. _____
   E. _____
   F. _____

6. What AWS brazing flux type number is used for brazing a nickel or nickel-based alloy?
   _____

7. What form of flux is used when brazing magnesium alloys?
   _____

8. Brazing can be used to combine two different metals. List the three brazing filler metals that can be used when brazing cast iron and copper?
   _____
   _____
   _____

9. List eight criteria that should be considered when choosing a brazing flux.
   _____
   _____
   _____
   _____
   _____
   _____
   _____
   _____

10. *True or False?* Braze welding fluxes must withstand higher temperatures for longer periods than brazing fluxes.

    10. _____

256   Modern Welding Lab Workbook

Job 16A-1

Braze Welding a Butt Joint, a Lap Joint, and a T-Joint in the Flat Welding Position

Name _____ Date _____

Class _____ Instructor _____

Learning Objective
- In this job, you will braze weld butt joints, lap joints, and T-joints in the flat welding position.

1. Obtain six pieces of mild steel that measure 1/8″ × 1 1/2″ × 5″ (3.2mm × 40mm × 125mm).

2. Clean the joint areas of all pieces at least 1/2″ (12.5mm) back from the joint.

3. What flux is recommended? (Refer to Figure 16-10 in the text.)

   _____

4. Braze weld two of each of the following joints.

   Note: Tack braze each joint at three points to hold the joint in position.

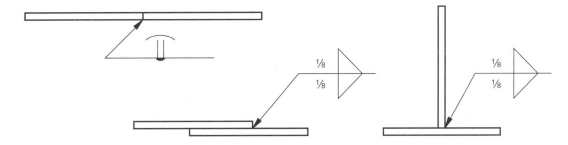

Inspection
The braze-welded beads should be straight, with an even bead width.

Instructor's initials: _____

Lesson 16B

Brazing and Braze Welding Processes

Name _____ Date _____

Class _____ Instructor _____

Learning Objective
- You will be able to braze stainless steel, cast iron, and some nonferrous metals. You will also be able to determine which flux to use for each metal.

Instructions
Carefully read Headings 16.6 through 16.8 of the text. Also study Figures 16-23 through 16-41 in the text. Then answer the following questions.

1. List five sources of heat for brazing.

   _____

   _____

   _____

   _____

   _____

2. List three factors that you should consider when selecting a brazing filler metal.

   _____

   _____

   _____

3. List the four brazing alloys suggested for brazing copper to copper. (See Figure 16-15 in the text.)

   _____

   _____

   _____

   _____

4. *True or False?* Brazing molybdenum (Mo) to nickel (Ni) is *not* recommended.   4. _____

Copyright by The Goodheart-Willcox Co., Inc.   Modern Welding Lab Workbook   259

5. List the five brazing filler metals recommended for joining tool steel to carbon steel.
   _____
   _____
   _____
   _____

6. A silver-brazed joint is strongest when the thickness of the silver brazing filler metal in the joint is _____.
   A. .002"
   B. .006"
   C. .009"
   D. .012"

   6. _____

7. Brazing filler metal will not flow over the base metal surface unless the base metal surface is heated to the brazing filler metal _____ temperature.

   7. _____

8. In silver brazing, the filler metal should be added when the flux appears _____.
   A. clear
   B. milky
   C. dry
   D. The flux appearance does not indicate when filler metal should be added.

   8. _____

9. Which of the following is *not* a method used to clean magnesium after brazing is done?
   A. Cleaning in hot running water.
   B. Scrubbing with an alkaline solution.
   C. Mechanical scrubbing.
   D. Dipping into a chrome pickling solution.

   9. _____

10. For brazing cast iron, _____ is recommended.
    A. a preheating temperature of 400°F to 600°F (204°C to 316°C)
    B. tinning of the surfaces to be brazed
    C. a tip that provides high heat with high gas pressures
    D. Both A and B.

    10. _____

260 Modern Welding Lab Workbook

Job 16B-1

Braze Welding a V-Groove on an Outside Corner in the Flat and Horizontal Welding Positions

Name _____ Date _____

Class _____ Instructor _____

Learning Objective
- In this job, you will braze weld a V-groove on an outside corner joint in the flat and horizontal welding positions.

1. Obtain three pieces of carbon steel that measure 1/4" × 1 1/2" × 5" (6.4mm × 40mm × 125mm).

2. Braze weld the pieces into the shape shown in the following drawing.
    - Begin by braze welding a V-groove outside corner joint on the first two pieces in the flat position.
    - Next, tack weld the third piece in place so the weldment looks like the shape shown below.
    - Braze weld the second V-groove outside corner joint in the horizontal position.

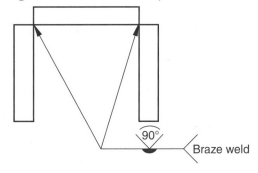

3. If no silver-, gold-, nickel-, or zinc-based filler metal is available, what filler metal would you use? (See Figure 16-15 in the text.)

_____

Inspection
The braze welding bead should be convex. The bead should not go beyond the edges of the groove more than about 1/16" (1.6mm). The ripples in the bead should be evenly spaced, and the bead should have a constant width. There should be complete penetration over the entire length of the joint.

Instructor's initials: _____

Lesson 17A
Resistance Welding Safety

Name _____ Date _____

Class _____ Instructor _____

Learning Objective
- You will be able to use resistance welding equipment safely.

Instructions
Carefully read Headings 17.13, 18.2.2, 18.2.3, and 18.8 in the text. Then answer the following questions.

1. When molten metal squirts out from a spot weld, the weld is called a(n) ____ weld.
   1. _____

2. ____ should be worn to prevent injuries resulting from flying molten metal and flying sparks.
   A. Safety glasses
   B. Flash goggles or a face shield
   C. Gloves
   D. Protective clothing
   E. All of the above.
   2. _____

3. What do proper lock-out, tag-out procedures consist of?
   _____
   _____
   _____

4. You should always wear ____ when handling spot welded metal and sheet metal.
   4. _____

5. List two hazards that can result from an expulsion spot weld. _____
   _____
   _____

Copyright by The Goodheart-Willcox Co., Inc.   Modern Welding Lab Workbook   263

6. Which of the following is *not* true of resistance welding equipment?
   A. The welding voltage across the electrodes is very high.
   B. The primary circuit wiring should be handled only by qualified electricians.
   C. All resistance welding equipment should be grounded.
   D. The secondary current is very high.

6. _____

7. Why are dual palm switches used on some resistance welding machines?

_____

_____

_____

8. Which of the following should be done prior to starting work on a welding machine secondary? (More than one answer may be correct.)
   A. Turn off the resistance welding machine.
   B. Turn off the power at the substation.
   C. Turn off the machine's circuit breaker.
   D. Remove the electrodes from the machine.

8. _____

9. What should be done before work is started on a primary?

_____

_____

_____

10. If a hydraulic leak develops, you should ____.
    A. examine the machine to determine where the leak is
    B. tell everyone in the class of the problem
    C. turn off the machine and tell your instructor or supervisor
    D. repair the leak with black electrical tape

10. _____

Lesson 17B

Resistance Welding Machines

Name _____ Date _____

Class _____ Instructor _____

Learning Objective
● You will be able to describe the types of resistance welding machines, their major components, and the types of welds produced by resistance welding.

Instructions
Carefully read Headings 17.1 to 17.4 and 17.8 to 17.13 of the text. Review Headings 4.4 through 4.4.4 of the text. Also, study the Figures referred to under these Headings. Then answer the following questions.

1. List the five main parts of a resistance welding machine.
   _____
   _____

2. *True or False?* The secondary windings of a resistance welding machine are usually water-cooled, but small transformers can be air-cooled.

    2. _____

3. A transformer capacity is listed as a _____ rating.
   A. VA
   B. KVA
   C. percent heat
   D. duty cycle

    3. _____

4. Label the parts of the transformer shown in the following image.
   A. _____
   B. _____
   C. _____
   D. _____

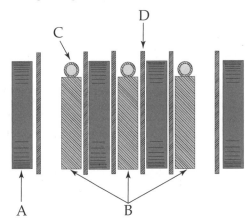

Copyright by The Goodheart-Willcox Co., Inc.   Modern Welding Lab Workbook   **265**

5. The term _____ is used to describe the amount of time a transformer delivers current in ratio to the time the current is off.
   A. KVA
   B. percent heat
   C. duty cycle
   D. tap setting

   5. _____

6. A resistance welding transformer is usually rated at _____ percent duty cycle. The rating is based on a(n) _____ minute time period.

   6. _____
   _____

7. A transformer that increases the supplied current and decreases the supplied voltage is called a(n) _____ transformer.

   7. _____

8. Which of the following methods will change the force applied by a pneumatic pressure system?
   A. Set desired value on robot controller.
   B. Change the mechanical leverage.
   C. Adjust the regulator.
   D. All of the above.

   8. _____

9. What three features should be considered when a resistance welding machine is selected?
   _____
   _____
   _____

10. List the three ways a resistance welding machine can obtain electrical energy.
    _____
    _____
    _____

11. A three-phase machine has _____ SCRs and _____ primary transformers. (Fill in each of the blanks with a number.)

    11. _____
    _____

12. Making multiple projection welds requires _____ compared to making single projection welds.
    A. higher forces and lower currents
    B. the same forces but higher currents
    C. higher forces and higher currents
    D. lesser forces and lower currents

    12. _____

13. List the two electrode drive mechanisms used on seam welding machines.
    _____

14. *True or False?* A seam welding machine that has the wheel electrodes perpendicular to the front of the machine will produce a transverse seam.

    14. _____

Lesson 17B Resistance Welding Machines

Name _____

15. See the following drawing. The dimension marked *A* is called the ____ and determines the maximum ____ of a part that can be welded.

    The dimension marked *B* is called the ____ and determines the maximum ____ of a part that can be welded.

15. _____

    _____

    _____

    _____

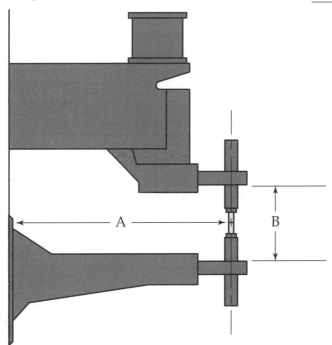

16. Complete the following sketch to show the appearance of a completed upset weld and a completed flash weld.

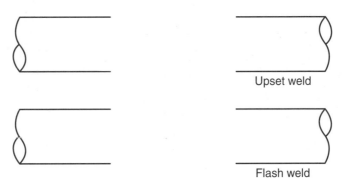

Upset weld

Flash weld

17. Which process does not require the parts to be cleaned, flash welding or upset welding?

17. _____

18. *True or False?* An upset welding machine uses the same type of transformer as a spot welding machine.

18. _____

19. What type of machine is shown in Question 20?

_____

Copyright by The Goodheart-Willcox Co., Inc.   Modern Welding Lab Workbook   267

20. Identify the parts of the machine indicated in the following drawing.

A. _____
B. _____
C. _____
D. _____
E. _____
F. _____
G. _____
H. _____

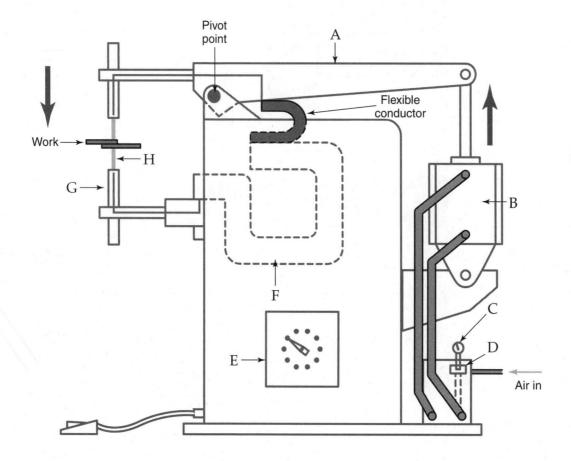

Job 17B-1

Major Components of Resistance Spot Welding Machines

Name _____ Date _____

Class _____ Instructor _____

> **Learning Objective**
>
> • In this job, you will identify the major components of the resistance spot welding machines in your shop. Your instructor will select a machine to use for this and the following assignment. This job assignment will be done as an instructor demonstration with student participation. A group of students may work together. **The machine will remain off during this assignment.**

1. Read Heading 17.13 of the text and listen to the safety instructions given by your instructor. Check to make sure the machine is off. Your instructor may also remove the fuses or turn off the circuit breaker for added safety.

2. Who is the manufacturer of the machine you are studying?
   _____

3. What is the machine's KVA rating? (See the nameplate.)
   _____

4. What type of machine is it? 4. _____
   A. Rocker-arm
   B. Press
   C. Portable
   D. Seam
   E. Other

5. Your instructor will open the controller cabinet or the side of 5. _____
   the machine and point out the main power connections. *Never*
   open the controller cabinet or side when the power is on. How
   many SCRs are there?

6. How does the machine obtain its electrical energy? 6. _____
   A. Single-phase machine
   B. Three-phase machine
   C. Stored energy machine

7. What is the current capacity of the SCRs used in the machine? 7. _____

8. Your instructor will point out the transformer on the machine and show you the main power connection to the transformer. Measure the height, length, and width of the transformer.

   Height: _____

   Length: _____

   Width: _____

9. Measure the throat depth and horn spacing of the machine.

   Throat depth: _____

   Horn spacing: _____

10. Can either of the dimensions listed in Question 9 be changed? If so, how? If not, why not?

    _____

    _____

    _____

11. Your instructor will point out the safety features on the machine. After discussion of the parts within the control cabinet, your instructor will close the control cabinet or side of the machine.

Inspection

You and your instructor should check to make sure the machine is in good, safe working condition before you leave it.

Instructor's initials: _____

Lesson 17C

Resistance Welding Electrical Components

Name _____ Date _____

Class _____ Instructor _____

Learning Objective
- You will be able to describe the electrical components of a resistance welding machine, including the electrodes and electrode holders. You will also be able to explain how the cooling system of a resistance welding machine operates.

Instructions
Carefully read Headings 17.5 to 17.7 of the text. Also study Figures 17-14 through 17-26 in the text. Then answer the following questions.

1. List the three times that are used to control a resistance spot weld.
   _____

2. Label the parts on the SCR shown and draw an arrow to indicate the direction of the current flow.

   A. _____
   B. _____ (+ or -)
   C. _____
   D. _____ (+ or -)
   E. _____

3. *True or False?* When the controller sends a signal to the gate side of an SCR, the SCR stops conducting current.

   3. _____

4. An increase in _____ will *not* produce a higher welding current.
   A. welding time
   B. tap setting
   C. percent heat
   D. current set value

   4. _____

5. Each SCR will conduct electricity _____ times per second.
   A. 30
   B. 60
   C. 90
   D. 120

   5. _____

Copyright by The Goodheart-Willcox Co., Inc.

6. Label the parts of the resistance welding circuit shown.

   A. _____
   B. _____
   C. _____
   D. _____
   E. _____
   F. _____
   G. _____
   H. _____

7. List four requirements of resistance welding electrodes.

   _____
   _____
   _____
   _____

8. What group and class of electrode is used for high production spot welding of mild steels and stainless steels?

   _____

9. Label the parts of the electrode shown. Use arrows to show the direction of waterflow through the electrode.

   A. _____
   B. _____

10. List the three ways an electrode or adaptor can be attached to an electrode holder.

    _____
    _____
    _____

272   Modern Welding Lab Workbook

Job 17C-1

Resistance Spot Welding Machine Cooling and Electrodes

Name _____   Date _____

Class _____   Instructor _____

Learning Objective

- In this job, you will examine the role of water cooling in a resistance spot welding machine. You will also study the construction of electrodes. Your instructor will select a machine on which to perform this assignment. **The machine will remain off during this assignment.**

1. Follow your instructor's safety instructions. Do not turn the machine on during this assignment. Your instructor will work with you while you are performing this job assignment. Students can work in groups.

2. List the manufacturer and KVA rating of the machine.

   _____

   _____

3. On the nameplate of most resistance welding machines, specific information is given regarding minimum water pressure and maximum water temperature. Find this information and copy it on the lines provided.

   Note: If the information is not available, write "not available."

   _____

   _____

   _____

4. List the parts of a resistance welding machine that are usually water-cooled.

   _____

   _____

   _____

5. Locate the water inlet on the machine. From this starting point, follow the flow direction of the cooling water through the machine by tracing the hoses. After cooling the machine parts, the water leaves the machine. In the space provided on the following page, sketch the path the cooling water followed in your machine. Sketch the parts of the machine that are cooled and any valves and/or branches along the path of the water hoses.

Copyright by The Goodheart-Willcox Co., Inc.   Modern Welding Lab Workbook   273

WATER INLET

WATER OUTLET

6. Are all of the machine parts listed in Question 4 water-cooled   6. _____
in the machine you examined?

If not, list those parts not water-cooled. _____

_____

7. Turn off the water to the machine. Remove one of the electrodes from the electrode holder. If your machine has an adaptor and cap, take them apart. Have a large cup and a towel available to catch any water that leaks when the electrode is removed. In the space provided, sketch the electrode you removed from your machine. Measure the length, the diameter, the tip face diameter, and any other important features. Write the dimensions on your sketch. If the electrode has a cap and an adaptor, sketch and dimension both.

274   Modern Welding Lab Workbook

Job 17C-1 Resistance Spot Welding Machine Cooling and Electrodes

Name _____

8. What type of electrode face does the electrode have? _____
_____

9. To determine how close to the end of the electrode the cooling water flows, do the following.

   A. Measure the length of the electrode or the cap.   9. A. _____

   B. Measure the depth of the hole where the water flows.   B. _____

   C. How close to the end of the electrode does the cooling   C. _____
      water flow? (Subtract line B from line A to get the answer.)

■ **Note:** If the hole is deep, insert a piece of 0.035″ or 0.045″ wire down to the bottom of the hole. Mark the wire even with the top of the electrode. Remove the wire and measure from the mark to the end of the wire. Enter this value in line B. Subtract the depth of the hole (line B) from the length of the electrode or cap (line A). This is how close the water comes to the end of the electrode.

Instructor's initials: _____

Lesson 18
Resistance Welding

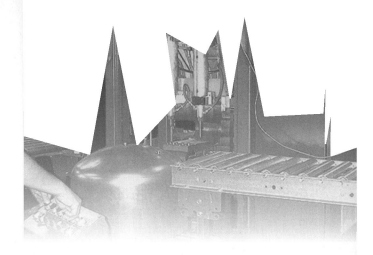

Name _____ Date _____
Class _____ Instructor _____

Learning Objective
You will be able to determine the proper settings for resistance welding mild steel. You will be able to set up the machine for welding. You will also be able to produce spot welds.

Instructions
Carefully read Headings 17.4 and 18.2 through 18.2.5 of the text. Also study Figures 17-11 through 17-13 and Figures 18-2 through 18-21 in the text. Then answer the following questions.

1. Define the following terms.

   A. Weld time _____

   B. Hold time _____

   C. Squeeze time _____

   D. Off time _____

2. Give a brief definition of the following terms.

   A. Force gauge _____

   B. Current analyzer _____

   C. Controller _____

3. *True or False?* The tap switch on the transformer is used to make large changes in the secondary current.   3. _____

Copyright by The Goodheart-Willcox Co., Inc.   Modern Welding Lab Workbook   277

4. The _____ is used to make small changes in current.   4. _____
   A. duty cycle
   B. KVA rating
   C. percent heat control
   D. weld time

5. On the following weld schedule, identify the axes marked A, B, and C and the areas marked D, E, and F.

   A. _____
   B. _____
   C. _____
   D. _____
   E. _____
   F. _____

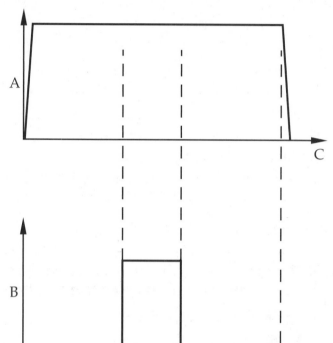

6. Develop a weld schedule to weld one piece of 0.060" (1.5mm) thick mild steel and one piece of 0.048" (1.2mm) thick mild steel. (See Heading 18.2.2 and Figure 18-7 in the text.)

   A. What weld time should be used? _____
   B. What welding current should be used? _____
   C. What electrode force should be used? _____
   D. What size should the complete weld be? _____

Lesson 18 Resistance Welding

Name _____

7. Fill in the following graph and draw a weld schedule using answers from Question 6. Use a 20 cycle squeeze time and a 30 cycle hold time. Identify the parts of the graph.

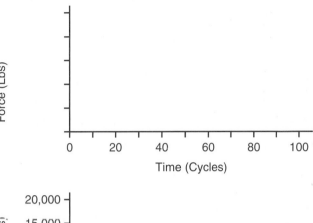

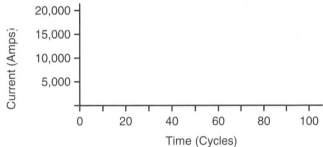

8. How is the desired electrode force set when using an air-operated (pneumatic) force system?
_____

9. After welding hundreds or thousands of spot welds on mild steel, the electrodes begin to _____. Reshaping an electrode is also referred to as _____ the electrode.

9. _____
   _____

10. Where does a high-frequency current flow in a conductor?
_____

Job 18-1

Setting the Variables and Making Spot Welds

Name _____ Date _____

Class _____ Instructor _____

> **Learning Objective**
> - In this job, you will set the variables required to make a high-quality spot weld. You will also make and test several spot welds. Your instructor will assign a machine to you for this assignment.

Who manufactures the machine you are using? _____

_____

Obtain ten pieces of mild steel about 2" × 8" (50mm × 200mm). All pieces should be the same thickness, but they do not have to be. On each set of two pieces, 6 to 8 spot welds will be made. If the metal is rusty or has grease, oil, or paint on it, it must be cleaned. Use a degreasing solution and/or a wire brush to remove all of these from the metal.

1. How thick is the metal you will be welding? _____

2. From Figure 18-7 in the text, determine the following:
    A. What tip diameter should be used? _____
    B. What weld time should be used? _____
    C. What weld current should be used? _____
    D. What weld electrode force should be used? _____
    E. What weld size should be produced? _____

3. Select an electrode or electrode cap close to the size calculated in Question 2 above.
    A. What is the actual diameter? _____
    B. What face design does the electrode have? _____

4. Install or attach the electrode or electrode cap to the welding machine. (See Heading 18.2.3 of the text.)

5. Turn on the welding machine. Turn on the water. Find the switch marked *Weld–No Weld* and place it in the *No Weld* position. This allows all the functions of the machine to operate, but no electrical current will flow through the electrode.

6. Set a 30 to 60 cycle squeeze time, 0 cycles weld time, and a 30 cycle hold time on the controller. If a force gauge is available, use it to set the correct electrode force. Adjust the regulator on a pneumatic or hydraulic machine to obtain the correct force. Adjust the spring tension on a manual force machine to obtain the correct force.

7. If a force gauge is not available, use the following technique to set the pressure.
   A. Place two pieces of the material to be welded between the electrodes.
   B. Press the foot control or palm switch to make the electrodes close on the metal. If no indentation is made on the metal, increase the electrode force by increasing the pneumatic or hydraulic pressure, or by increasing the spring tension.
   C. Continue increasing the electrode force and cycling through additional weld sequences until the electrodes indent the metal slightly.
   D. Next, reduce the electrode force slowly until no indentation occurs. (If this force still seems to be high, reduce the force slightly.)

8. The air pressure or hydraulic pressure is set at _____ psig. (No answer is required if a spring force machine is used.)

   8. _____

9. Figure 18-7 in the text lists the minimum distance that must be maintained between spot welds based on metal thickness. What weld spacing should be used for the thickness of metal you are using?

   9. _____

10. On the controller, set the weld time determined in Question 2. Switch the *Weld–No Weld* switch to *Weld*. Set the weld current. Use the methods described in Step 10 or Step 12.

11. To determine the current setting by the trial-and-error method, do the following:
    A. Set the tap switch to a value less than half the highest value. If there are eight tap settings, select number 3. If there are six tap settings, select number 2. If there are five or fewer settings, select number 1.
    B. Set the percent heat control at 70%.
    C. Place two pieces of metal to be welded between the electrodes.
    D. Press the foot control or palm switch to begin the welding sequence.
    E. If an expulsion weld occurs, reduce the tap setting by two. If the two pieces are not welded at all, or if they can be torn apart by hand, increase the tap setting by one.
    F. Provided neither of the two conditions in Step E exists, the weld must be torn apart to examine the weld size. The technique for peel testing a weld is described at the end of this job assignment. Follow the steps described there and measure the weld nugget size. Make changes to obtain the nugget size listed in the answer to Question 2E on the previous page. If the weld nugget is small, increase the current by increasing the tap setting or the percent heat control.
    G. Continue welding additional welds following Steps D, E, and F until the weld size is within 10% of the target diameter. Each spot weld will vary slightly in size, so weld three or more samples to make sure the size remains nearly the same.

    ■ **Note:** If the percent heat control is at 90% or above, and you need just a little more current, increase the tap setting by one number and reduce the percent heat control to 50%. If the percent heat control is at 40% or less, and you need slightly less current, decrease the tap setting by one and increase the percent heat control to 85%.

12. To set the current using a current analyzer, do the following:
    A. Set the tap switch to a value less than half the highest value. If there are eight tap settings, select number 3. If there are six tap settings, select number 2. If there are five or fewer settings, select number 1.
    B. Set the percent heat control at 70%.
    C. Place the pick-up coil around the lower electrode and turn on the current analyzer.
    D. Place two pieces of metal to be welded between the electrodes.
    E. Press the foot control or palm switch to begin the welding sequence.
    F. The welding current measured in amps or kiloamps (1000 amps) will be displayed on the current analyzer.

Job 18-1 Setting the Variables and Making Spot Welds

Name _____

G. Adjust the tap setting and percent heat control to obtain the desired current. If the value on the current analyzer is too low, increase the tap setting or percent heat control. If the value is too high, decrease the tap setting or percent heat control.

■ **Note:** If the percent heat control is at 90% or above, and you need just a little more current, increase the tap setting by one number and reduce the percent heat control to 50%. If the percent heat control is at 40% or less, and you need slightly less current, decrease the tap setting by one and increase the percent heat control to 85%.

H. Repeat Steps E through G until the current analyzer reads the correct value.
I. Once the correct current is set, make a spot weld and peel test it. Check the diameter of the weld to make sure the set current will produce a weld of proper size. If the weld size is not correct, follow Steps J through L to adjust the settings to produce the desired weld size.
J. Press the foot control or palm switch to begin the welding sequence and produce a weld.
K. Peel test the weld and measure the nugget diameter. Increase current or weld time to increase the nugget size. Decrease current or weld time if there is expulsion.
L. Continue welding additional welds following Steps J and K until the weld size is within 10% of the target diameter. Each spot weld will vary slightly in size, so weld three or more samples to make sure the size remains nearly the same.

13. What tap setting is used? 13. _____

    The heat control is set at _____ %.   _____

14. Your machine should now be properly set to weld.

15. Take two pieces of 2″ × 8″ (50mm × 200mm) mild steel and overlap them about 7 1/2″ (190mm). Start 1 1/2″ (40mm) from one end. Make as many welds as possible down the center of the two pieces, maintaining the minimum weld spacing from Step 9.

16. Weld three sets of material together. Tear them apart using the techniques described at the end of this job.

17. Measure the sizes of two welds from each piece and list the sizes on the following lines.

    ■ **Note:** Notice the variation even when no change is made in the machine settings.

    Two welds from piece #1 _____
    Two welds from piece #2 _____
    Two welds from piece #3 _____

18. In the next few steps, you will study the effects of weld spacing. Previously, welds were made using the minimum spacing shown in the chart in Figure 18-7 in the text. Welds made for the following task will be made at one-half or less of the spacing shown in the chart.

19. What is one-half of the minimum weld spacing shown in the   19. _____
    chart for the thickness of the material you are using?

20. Using the same machine settings as before, weld two pieces of material together with 12 spot welds. The first weld should be 1 1/2″ (40mm) from one end. The weld spacing should be equal to or less than the spacing from Step 19. Number the welds. The first weld is #1, the second is #2, etc. Tear the two pieces of metal apart.

21. Measure the diameter of each spot weld and record the size.

    #1 _____   #5 _____   #9 _____
    #2 _____   #6 _____   #10 _____
    #3 _____   #7 _____   #11 _____
    #4 _____   #8 _____   #12 _____

22. Does the weld size increase, decrease, or remain the same as each weld is made? _____   22. _____

    ■ **Note:** To prevent this from happening when spot welding, always maintain the minimum weld spacing. The weld size also changes when the welding electrodes begin to wear. To prevent this, the electrodes must be dressed regularly.

Inspection

Peel-Testing Spot Welds

A peel test separates the two pieces of metal that were spot welded together. The spot weld usually tears out one piece and can be measured on the other piece. (This does not always occur on thicker pieces of metal.)

Follow these steps to peel test a spot weld:

1. Bend back the 1 1/2" (40mm) portion of the metal that is not welded. One piece of metal will go 90° one way. The other piece will be bent 90° the other way. These can be pulled apart to examine the spot weld. See the following sketch and Figure 30-30 in the text.

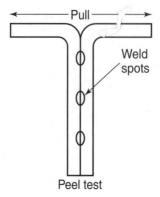

2. Place each bent piece in one jaw of a tensile test machine. Tighten each jaw onto the metal.

3. Apply a force to pull the pieces apart.

4. When the spot welds tear out of one piece of metal, stop pulling and remove each piece of metal.

5. With a set of calipers, measure the diameter of the spot welds. If no calipers are available, use a scale and measure the diameter. List the diameters of the spot welds.   5. _____

Instructor's initials: _____

284   Modern Welding Lab Workbook

Job 18-2

Setting the Variables and Making Spot Welds

Name _____   Date _____

Class _____   Instructor _____

> **Learning Objective**
>
> ● In this job, you will set the variables required to make a high-quality spot weld. You will also make and test several spot welds. This assigned job is the same as Job 18-1, except that you will use a different thickness of material or a different machine. All instructions are the same as in Job 18-1.

Who manufactures the machine you are using? _____
_____

Obtain ten pieces of mild steel about 2″ × 8″ (50mm × 200mm). All pieces should be ⟨the⟩ same thickness, but they do not have to be. On each set of two pieces, 6 to 8 spot welds will be made. If the metal is rusty or has grease, oil, or paint on it, it must be cleaned. Use a degreasing solution and/or a wire ⟨bru⟩sh to remove all of these from the metal.

1. How thick is the metal you will be welding? _____

2. From Figure 18-7 in the text, determine the following:
    A. What tip diameter should be used? _____
    B. What weld time should be used? _____
    C. What weld current should be used? _____
    D. What weld electrode force should be used? _____
    E. What weld size should be produced? _____

3. Select an electrode or electrode cap close to the size calculated in Question 2 above.
    A. What is the actual diameter? _____
    B. What face design does the electrode have? _____

4. Install or attach the electrode or electrode cap to the welding machine. (See Heading 18.2.3 of the text.)

5. Turn on the welding machine. Turn on the water. Find the switch marked *Weld–No Weld* and place it in the *No Weld* position. This allows all the functions of the machine to operate, but no electrical current will flow through the electrode.

6. Set a 30 to 60 cycle squeeze time, 0 cycles weld time, and a 30 cycle hold time on the controller. If a force gauge is available, use it to set the correct electrode force. Adjust the regulator on a pneumatic or hydraulic machine to obtain the correct force. Adjust the spring tension on a manual force machine to obtain the correct force.

7. If a force gauge is not available, use the following technique to set the pressure.
   A. Place two pieces of the material to be welded between the electrodes.
   B. Press the foot control or palm switch to make the electrodes close on the metal. If no indentation is made on the metal, increase the electrode force by increasing the pneumatic or hydraulic pressure, or by increasing the spring tension.
   C. Continue increasing the electrode force and cycling through additional weld sequences until the electrodes indent the metal slightly.
   D. Next, reduce the electrode force slowly until no indentation occurs. (If this force still seems to be high, reduce the force slightly.)

8. The air pressure or hydraulic pressure is set at _____ psig. (No answer is required if a spring force machine is used.)

   8. _____

9. Figure 18-7 in the text lists the minimum distance that must be maintained between spot welds based on metal thickness. What weld spacing should be used for the thickness of metal you are using?

   9. _____

10. On the controller, set the weld time determined in Question 2. Switch the *Weld–No Weld* switch to *Weld*. Set the weld current. Use the methods described in Step 10 or Step 12.

11. To determine the current setting by the trial-and-error method, do the following:
    A. Set the tap switch to a value less than half the highest value. If there are eight tap settings, select number 3. If there are six tap settings, select number 2. If there are five or fewer settings, select number 1.
    B. Set the percent heat control at 70%.
    C. Place two pieces of metal to be welded between the electrodes.
    D. Press the foot control or palm switch to begin the welding sequence.
    E. If an expulsion weld occurs, reduce the tap setting by two. If the two pieces are not welded at all, or if they can be torn apart by hand, increase the tap setting by one.
    F. Provided neither of the two conditions in Step E exists, the weld must be torn apart to examine the weld size. The technique for peel testing a weld is described at the end of this job assignment. Follow the steps described there and measure the weld nugget size. Make changes to obtain the nugget size listed in the answer to Question 2E on the previous page. If the weld nugget is small, increase the current by increasing the tap setting or the percent heat control.
    G. Continue welding additional welds following Steps D, E, and F until the weld size is within 10% of the target diameter. Each spot weld will vary slightly in size, so weld three or more samples to make sure the size remains nearly the same.

    ■ **Note:** If the percent heat control is at 90% or above, and you need just a little more current, increase the tap setting by one number and reduce the percent heat control to 50%. If the percent heat control is at 40% or less, and you need slightly less current, decrease the tap setting by one and increase the percent heat control to 85%.

12. To set the current using a current analyzer, do the following:
    A. Set the tap switch to a value less than half the highest value. If there are eight tap settings, select number 3. If there are six tap settings, select number 2. If there are five or fewer settings, select number 1.
    B. Set the percent heat control at 70%.
    C. Place the pick-up coil around the lower electrode and turn on the current analyzer.
    D. Place two pieces of metal to be welded between the electrodes.
    E. Press the foot control or palm switch to begin the welding sequence.
    F. The welding current measured in amps or kiloamps (1000 amps) will be displayed on the current analyzer.

Job 18-2 Setting the Variables and Making Spot Welds

Name _____

    G. Adjust the tap setting and percent heat control to obtain the desired current. If the value on the current analyzer is too low, increase the tap setting or percent heat control. If the value is too high, decrease the tap setting or percent heat control.

    ■ **Note:** If the percent heat control is at 90% or above, and you need just a little more current, increase the tap setting by one number and reduce the percent heat control to 50%. If the percent heat control is at 40% or less, and you need slightly less current, decrease the tap setting by one and increase the percent heat control to 85%.

    H. Repeat Steps E through G until the current analyzer reads the correct value.
    I. Once the correct current is set, make a spot weld and peel test it. Check the diameter of the weld to make sure the set current will produce a weld of proper size. If the weld size is not correct, follow Steps J through L to adjust the settings to produce the desired weld size.
    J. Press the foot control or palm switch to begin the welding sequence and produce a weld.
    K. Peel test the weld and measure the nugget diameter. Increase current or weld time to increase the nugget size. Decrease current or weld time if there is expulsion.
    L. Continue welding additional welds following Steps J and K until the weld size is within 10% of the target diameter. Each spot weld will vary slightly in size, so weld three or more samples to make sure the size remains nearly the same.

13. What tap setting is used?   13. _____

    The heat control is set at _____ %.   _____

14. Your machine should now be properly set to weld.

15. Take two pieces of 2″ × 8″ (50mm × 200mm) mild steel and overlap them about 7 1/2″ (190mm). Start 1 1/2″ (40mm) from one end. Make as many welds as possible down the center of the two pieces, maintaining the minimum weld spacing from Step 9.

16. Weld three sets of material together. Tear them apart using the techniques described at the end of this job.

17. Measure the sizes of two welds from each piece and list the sizes on the following lines.

    ■ **Note:** Notice the variation even when no change is made in the machine settings.

    Two welds from piece #1 _____
    Two welds from piece #2 _____
    Two welds from piece #3 _____

18. In the next few steps, you will study the effects of weld spacing. Previously, welds were made using the minimum spacing shown in the chart in Figure 18-7 in the text. Welds made for the following task will be made at one-half or less of the spacing shown in the chart.

19. What is one-half of the minimum weld spacing shown in the   19. _____
    chart for the thickness of the material you are using?

20. Using the same machine settings as before, weld two pieces of material together with 12 spot welds. The first weld should be 1 1/2″ (40mm) from one end. The weld spacing should be equal to or less than the spacing from Step 19. Number the welds. The first weld is #1, the second is #2, etc. Tear the two pieces of metal apart.

21. Measure the diameter of each spot weld and record the size.

   #1 _____   #5 _____   #9 _____
   #2 _____   #6 _____   #10 _____
   #3 _____   #7 _____   #11 _____
   #4 _____   #8 _____   #12 _____

22. Does the weld size increase, decrease, or remain the same as each weld is made?

   22. _____

   ■ **Note:** To prevent this from happening when spot welding, always maintain the minimum weld spacing. The weld size also changes when the welding electrodes begin to wear. To prevent this, the electrodes must be dressed regularly.

Inspection

Peel-Testing Spot Welds

A peel test separates the two pieces of metal that were spot welded together. The spot weld usually tears out one piece and can be measured on the other piece. (This does not always occur on thicker pieces of metal.)

Follow these steps to peel test a spot weld:

1. Bend back the 1 1/2" (40mm) portion of the metal that is not welded. One piece of metal will go 90° one way. The other piece will be bent 90° the other way. These can be pulled apart to examine the spot weld. See the following sketch and Figure 30-30 in the text.

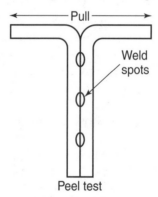

2. Place each bent piece in one jaw of a tensile test machine. Tighten each jaw onto the metal.

3. Apply a force to pull the pieces apart.

4. When the spot welds tear out of one piece of metal, stop pulling and remove each piece of metal.

5. With a set of calipers, measure the diameter of the spot welds. If no calipers are available, use a scale and estimate the diameter. List the diameters of the spot welds.

   5. _____
      _____
      _____
      _____

Instructor's initials: _____

Lesson 19A

Arc-Related Welding Processes

Name _____ Date _____

Class _____ Instructor _____

Learning Objective
- You will be able to describe special arc-related welding processes.

Instructions
Carefully read Headings 19.1 through 19.2.5 of the text. Also study Figures 19-1 through 19-20 in the text. Then answer the following questions.

1. Name the parts indicated on the following drawing of a submerged arc welding outfit.

   A. _____
   B. _____
   C. _____
   D. _____
   E. _____
   F. _____

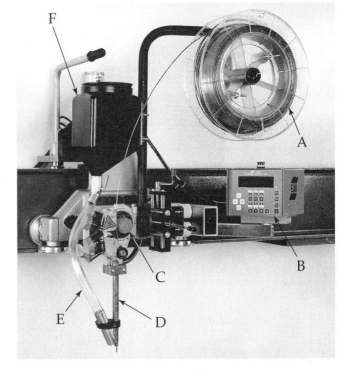

Modern Welding Lab Workbook   289

2. The welding operator must set the ____ when performing automatic submerged arc welding.
   A. voltage
   B. electrode extension
   C. wire feed speed
   D. travel speed
   E. alignment of the metal to be welded
   F. All of the above.

   2. _____

3. When multiple electrodes are used for submerged arc welding, which electrode combination is most common?
   A. Parallel connection transverse position.
   B. Series connection transverse position.
   C. Multiple power connection tandem position.
   D. Multiple power connection transverse position.

   3. _____

4. Electrogas welds are always made in the ____ welding position.

   4. _____

5. In electrogas welding, the electrodes may be ____ within the joint to ensure an equal distribution of heat from the arc(s).

   5. _____

6. Identify the parts of the electrogas welding process shown.
   A. _____
   B. _____
   C. _____
   D. _____
   E. _____
   F. _____
   G. _____
   H. _____
   I. _____
   J. _____
   K. _____

290   Modern Welding Lab Workbook

Name _____

7. What welding process is used to fill the weld joint when narrow gap welding is performed?
   _____

8. Which of the following is the most significant problem with narrow gap welding?
   A. Insufficient shielding gas to protect the molten metal.
   B. Inability to weld thick sections.
   C. Lack of penetration into the side walls.
   D. Consumption of the guide tube during the welding process.

   8. _____

9. *True or False?* Studs and fasteners can be welded to aluminum.

   9. _____

10. When large studs are welded using the arc stud welding process, a(n) _____ contains the molten metal during arcing.

    10. _____

11. The orifice gas in PAW cannot contain _____.
    A. oxygen
    B. nitrogen
    C. helium
    D. argon

    11. _____

12. Nontransferred PAW arc torches can be used to weld materials that do not conduct electricity.

    12. _____

13. What current and arc voltage should be used to weld 0.125″ (3.2mm) thick stainless steel using PAW? (Refer to Figure 19-20 in the text.)

    13. _____

14. *True or False?* Plasma arc welding uses a narrow column of plasma to produce a narrow weld with shallow penetration.

    14. _____

15. *True or False?* The plasma arc welding circuit shown in the following drawing is a nontransferred arc.

    15. _____

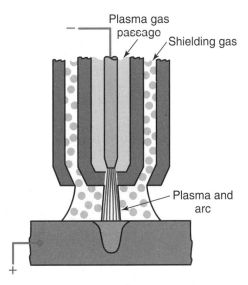

Lesson 19B

Solid-State and Other Welding Processes

Name _____ Date _____
Class _____ Instructor _____

Learning Objective
● You will be able to describe solid-state welding processes.

Instructions
Carefully read Headings 19.3 through 19.4.4 of the text. Also study Figures 19-21 through 19-54 in the text. Then answer the following questions.

1. *True or False?* During solid-state welding, the temperature of the material being welded is less than its melting point.

   1. _____

2. Match the following abbreviations with the correct welding process.

   _____ EXW   A. Friction stir welding
   _____ LBW   B. Diffusion welding
   _____ USW   C. Thermite welding
   _____ DFW   D. Forge welding
   _____ CEW   E. Friction welding
   _____ TW   F. Explosion welding
   _____ CW   G. Coextrusion welding
   _____ FRW   H. Laser beam welding
   _____ FSW   I. Cold welding
                     J. Ultrasonic welding

3. Cold welding is used on _____ metal.
   A. hard
   B. ductile
   C. brittle
   D. thick

   3. _____

4. What is the most common application of explosion welding?
   _____

5. *True or False?* Forge welding is usually limited to the welding of solid steel stock.

5. _____

6. Parts to be forge welded are heated to a(n) _____ heat in a(n) _____ forge.

6. _____

_____

7. *True or False?* In friction welding, heat is generated when one part rotates against a stationary (not rotating) part.

7. _____

8. Describe the rotating tool used in friction stir welding.

_____

_____

_____

9. List three base metals that can be welded by friction stir welding.

_____

_____

10. *True or False?* Ultrasonic welding equipment converts high-frequency electrical power into mechanical vibrations.

10. _____

11. List four advantages of ultrasonic welding.

_____

_____

_____

_____

_____

12. During electroslag welding, a layer of _____ covers the molten weld metal.

12. _____

13. Which element is more chemically active: aluminum or iron?

13. _____

14. After a thermite mixture is ignited, a temperature of _____°F (_____°C) will be reached.

14. _____

15. Why is the thermite welding process considered to be safe?

_____

_____

Lesson 19B Solid-State and Other Welding Processes

Name _____

16. Identify the basic parts shown in the following drawing of an electron beam welder.
    A. _____
    B. _____
    C. _____
    D. _____
    E. _____
    F. _____
    G. _____

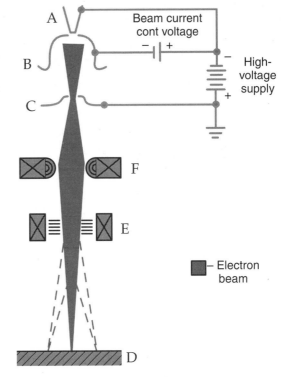

17. A complete penetration electron beam weld on ____" (____ mm) thick steel plate can be made in one pass.

17. _____

18. In electron beam welding, the voltage difference between the emitter and the anode is known as the ____ voltage.
    A. primary
    B. vacuum
    C. accelerating
    D. secondary

18. _____

19. A(n) ____ laser can be used to weld continuously.
    A. ruby
    B. Nd YAG
    C. CO_2
    D. All of the above.

19. _____

20. List the parts of the CO_2 laser shown in the following drawing.

A. _____
B. _____
C. _____
D. _____
E. _____
F. _____
G. _____
H. _____

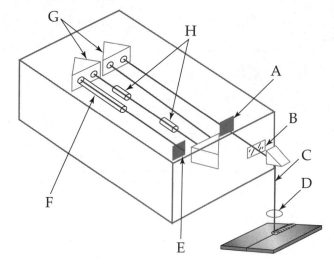

Lesson 20

Special Ferrous Welding Applications

Name _____ Date _____
Class _____ Instructor _____

Learning Objective
- You will be able to weld stainless steel, cast iron, and other ferrous alloys.

Instructions
Carefully read Chapter 20 of the text and study the figures in the chapter. Then answer the following questions.

1. The need for preheating medium- and high-carbon steels increases as the percentage of carbon in the steel increases.

    1. _____

2. What characteristic does manganese add to steel? _____
_____
_____

3. Which of the following alloying elements increases the toughness and strength of steel?
    A. Titanium
    B. Nickel
    C. Chromium
    D. Silicon

    3. _____

4. Preheating a medium- or high-carbon steel slows the rate of cooling, which helps reduce the formation of _____.

    4. _____

5. What can hydrogen in the weld area cause? _____
_____
_____

6. Low-alloys steels have less than _____ percent alloying elements.

    6. _____

Copyright by The Goodheart-Willcox Co., Inc. Modern Welding Lab Workbook   297

7. Preheating temperature for a 1″ thick AISI 4130 heat-treatable low-alloy steel is ____.
   A. 250°F (120°C)
   B. 400°F–550°F (200°C–290°C)
   C. 250°F–400°F (120°C–200°C)
   D. Preheating is not required.

7. _____

8. *True or False?* The two options for postweld heat treatment of low-alloy steels involve raising the temperature of the weldment immediately after welding.

8. _____

9. Should welding of an AISI 4130 heat-treatable low-alloy steel be done with a high or low heat input? Why?

_____

_____

10. ____ is *not* recommended for welding low-alloy steels.
    A. SMAW
    B. OFW
    C. GMAW
    D. FCAW

10. _____

11. List the *four* general classifications of stainless steel. _____

_____

_____

12. The "L" after an alloy designation means ____.

12. _____

13. Which of the following statements are true for martensitic stainless steel? (More than one answer may be correct.)
    A. It has good strength and is corrosion-resistant at elevated temperatures.
    B. It is often welded using austenitic (300 series) filler metal.
    C. It contains 15%–20% chromium.
    D. Preheating to 500°F (260°C) is required.

13. _____

14. Which of the following electrodes is recommended for gas metal arc welding of 304 stainless steel?
    A. E316
    B. ER310
    C. E308T-X
    D. ER308

14. _____

15. List a stainless steel electrode that meets each of the following descriptions.

    A. SMAW electrode _____

    B. Low-hydrogen SMAW electrode _____

    C. GMAW electrode _____

    D. Low-carbon GMAW electrode _____

Name _____

16. What type of shielding gases can be used for gas metal arc welding of stainless steel? _____
_____
_____

17. What methods can be used to reduce the grain growth that occurs when stainless steel is welded?
_____
_____
_____

18. *True or False?* To prevent porosity when welding stainless steel, a long arc length and a backing gas should be used.

18. _____

19. Maraging steel contains ____% carbon maximum, about ____% nickel, and about ____% cobalt.

19. _____
_____
_____

20. List the four main types of cast iron. _____
_____
_____
_____

21. Which SMAW electrode can be used to weld both gray and malleable cast iron?
    A. E70S-3
    B. ENiFeMn-CI
    C. ENi-CI
    D. ERNi-CI

21. _____

22. Why is cast iron preheated prior to welding? _____
_____

23. A. What preheat temperature should be used on gray cast iron prior to shielded metal arc welding?
_____

    B. What preheat temperature should be used on cast iron prior to oxyfuel gas welding?
_____

24. How is flux added to the weld when oxyfuel gas welding? _____
_____
_____

25. What postweld heat treatment is recommended for precipitation-hardening stainless steel?
_____
_____

Job 20-1

Square-Groove Weld on a Butt Joint on Stainless Steel Using SMAW in the Flat Welding Position

Name _____ Date _____

Class _____ Instructor _____

> **Learning Objective**
>
> In this job, you will make a square-groove weld on a butt joint on stainless steel in the flat welding position. You will use the SMAW process.

1. Obtain eight pieces of 304, 304L, 308, or 308L stainless steel. Other types may also be used. Each piece should measure 3/16″ × 2″ × 6″ (4.8mm × 50mm × 150mm). Also obtain four 1/8″ (3.2mm) diameter E308 or 308L electrodes. Your instructor will recommend other electrodes that can also be used.

2. Use a stainless steel wire brush to remove the oxides from the stainless steel plates. Clean along the 6″ (150mm) edges.

3. Set the machine for DCEP (DCRP).

4. The recommended current for each electrode can be obtained from the manufacturer. You can also use Figure 6-17 in the text. Use the E6010 column for E308 or E308L electrodes; however, the current must be further reduced by 10%–20%. The reduction adjusts for the differences in electrode material.

   What current should be used for a 1/8″ (3.2mm) 308 or 308L electrode? _____

   _____

5. Align two pieces of stainless steel to form a 6″ (150mm) long square-groove butt joint. Leave a 1/16″ to 1/8″ (1.6mm to 3.2mm) root gap.

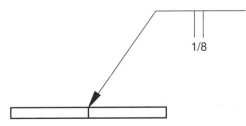

6. Tack weld each end and the middle of the joint.

7. Weld the joint in the flat welding position. Use the keyhole method to obtain complete penetration. Maintain a short arc length to prevent porosity and spatter.

8. Chip the slag, brush with a stainless steel wire brush, and examine the bead. Refer to Figure 6-23 in the text for examples of the effects of current, arc length, and travel speed on SMAW beads.

9. Make any corrections required, and weld three additional butt welds. You may want to attempt the final butt weld in the horizontal or vertical position.

Inspection

The beads should be straight and have complete penetration. The beads should have a uniform buildup, evenly spaced ripples, uniform penetration, and good fusion with no overlap or undercut. Save these welded samples for use in Job 20-2.

Instructor's initials: _____

Job 20-2

Fillet Weld on a T-Joint on Stainless Steel Using SMAW in the Flat Welding Position

Name _____ Date _____

Class _____ Instructor _____

Learning Objective

- In this job, you will make a fillet weld on a stainless steel T-joint in the flat or horizontal welding position using the SMAW process. You will be using the plates welded in Job 20-1 to create the T-joint.

1. Obtain four 1/8″ (3.2mm) diameter E308 or E308L electrodes and the samples that were welded in Job 20-1.

2. Before being welded, the plates must be cleaned again. Stainless steel wire brush an area down the length of two of the welded samples. Wire brush the edge of two other welded samples.

3. Align one clean edge onto the clean area of another plate to form a T-joint. Tack weld the ends and middle of the plates.

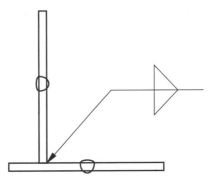

4. Weld a fillet weld in the flat or horizontal welding position.

5. Remove the slag and examine the weld. If the weld does not appear correct, discuss your technique with your instructor.

6. Weld the opposite side of the joint you have already welded. Then weld the other two plates together using a double fillet weld.

Inspection

Fillet welds in stainless steel should be very similar to fillet welds in mild steel. The surface of the bead should have regularly spaced ripples and an even buildup with no overlap or undercut.

Instructor's initials: _____

Job 20-3

Square-Groove Weld on a Butt Joint and a Fillet Weld in the Flat Welding Position

Name _____ Date _____

Class _____ Instructor _____

> **Learning Objective**
> - In this job, you will make square-groove welds on a butt joint and fillet welds on a T-joint. You will perform gas tungsten arc welding on stainless steel.

1. Obtain nine pieces of 304, 304L, 308, 308L, or another type of stainless steel as your instructor advises. Each piece should be 1/8″ × 2″ × 6″ (3.2mm × 50mm × 150mm). Obtain three 36″ (92cm) lengths of 3/32″ (2.4mm) diameter filler rod, ER308, ER308L, or another alloy as your instructor advises.

2. Use a 3/32″ (2.4mm) diameter thoriated tungsten electrode. Sharpen the electrode as discussed in Heading 8.3.5 and Figures 8-31 and 8-32 in the text.

3. A. What amperage should be used to butt weld 1/8″ (3.2mm) stainless steel? (Refer to Figure 8-22 in the text.)

   B. What amperage should be used to fillet weld 1/8″ (3.2mm) stainless steel?

   3. A. _____

   B. _____

4. Set the machine for DCEN (DCSP). Set the amperage on the machine. Also, adjust the argon flow rate to 15 cfh.

5. Using a stainless steel wire brush, clean the edges of the material to be welded.

6. Align two pieces to form a 6″ (150mm) square-groove butt weld. Leave a 1/16″ (1.6mm) gap. Tack weld the middle and both ends of the joint.

7. Weld the joint using the keyhole method. Full penetration can be achieved in one pass.

8. Examine the completed joint and discuss your results and techniques with your instructor.

9. Align and tack another piece to the pieces already welded. Complete the weld on the new butt joint. Add additional pieces until you have made the four butt joint welds shown.

10. Refer to the following figure to complete a welded assembly. Tack weld a piece per the drawing to create a T-joint.

Job 20-3 Square-Groove Weld on a Butt Joint and a Fillet Weld in the Flat Welding Position

Name _____

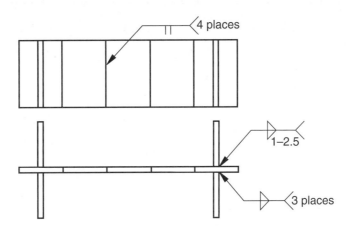

■ **Note:** One of the fillet welds has intermittent welds. This will allow you to practice your starts and stops. If you are having any difficulty while welding this assembly, talk to your instructor about the techniques required for welding stainless steel.

Inspection

Each fillet weld must be even in width, with evenly spaced ripples in the beads. The beads should be convex with no overlapping or undercutting. Each butt weld should have evenly spaced ripples, and a constant width and contour. The butt welds should also have full penetration over the entire length of the weld.

Instructor's initials: _____

Job 20-4

Square or V-Groove Weld on a Butt Joint on Cast Iron Using SMAW

Name _____ Date _____
Class _____ Instructor _____

> **Learning Objective**
>
> • In this job, you will weld a butt joint in cast iron using the SMAW process.

1. Obtain six pieces of cast iron measuring approximately 3/8″ × 2″ × 6″ (9.5mm × 50mm × 150mm).

   ■ **Note:** Any available size can be used.

2. Prepare the pieces for butt welding as described in Heading 20.9 and Figure 20-18 in the text. Prepare both long edges of each piece, so that both edges can be used. Bevel any pieces over 1/4″ (6.4mm) in thickness.

3. List three electrode types used to weld cast iron.

   _____

   _____

4. Obtain six 5/32″ (4mm) diameter electrodes used for welding 3/8″ (9.5mm) thick cast iron. If a different thickness of cast iron will be welded, refer to Figure 6-17 in the text to determine a proper electrode diameter.

5. Set the machine for DCEP (DCRP) or AC. Follow the electrode manufacturer's suggested current settings. The current used for welding cast iron is the same or slightly less than the current used for the same thickness of mild steel.

6. Preheat the cast iron to 500°F–600°F (260°C–315°C). Place the pieces in a preheated oven for one-half hour. If an oven is not available, use an oxyfuel gas torch and temperature-indicating crayons or liquid to determine the correct preheat temperature.

7. Align two pieces to form a 6″ (150mm) long butt joint. Leave a 1/16″ to 1/8″ (1.6mm to 3.2mm) root gap. Tack weld the pieces at each end and in the middle.

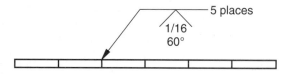

8. Weld the butt joint. The welding procedure is similar to welding mild steel. Use stringer beads and clean the slag between each pass.

9. Weld additional pieces of cast iron onto the initial two pieces until five groove welds are completed. Follow Steps 6, 7, and 8.

> **Inspection**
>
> Inspect the weld as you would a mild steel weld. The surface of the bead should have an even buildup and regularly spaced ripples. The root should have consistent penetration.

Instructor's initials: _____

Lesson 21
Nonferrous Welding Applications

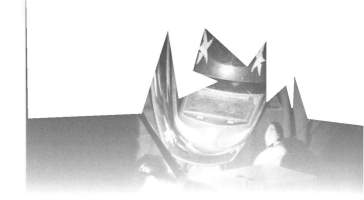

Name _____ Date _____

Class _____ Instructor _____

Learning Objective
You will be able to describe how to weld nonferrous metals, such as aluminum, magnesium, copper, and titanium, and their alloys. You will also be able to describe the welding of plastics.

Instructions
Carefully read Chapter 21 of the text and study the figures in the chapter. Then answer the following questions.

1. Which of the following are classified as nonferrous metals or alloys? (More than one answer may be correct.)
   A. Copper
   B. Aluminum
   C. Zirconium
   D. Cast iron
   E. Titanium
   F. Zinc
   G. Maraging steel

   1. _____

2. Match the aluminum alloy classification with its major alloying element.
   _____ 6061
   _____ 5056
   _____ 1100
   _____ 2024
   _____ 7075
   _____ 4043

   A. Silicon
   B. Magnesium and silicon
   C. 99% aluminum (minimum)
   D. Copper
   E. Other
   F. Zinc
   G. Manganese
   H. Magnesium

3. When aluminum is welded, a backing strip is often used to control the _____ and _____ of the penetration.

   3. _____

4. Which of the following is *not* a reason why aluminum is difficult to weld?
   A. Aluminum oxidizes at high temperatures.
   B. Aluminum oxide is heavier than aluminum.
   C. Aluminum reflects the heat of the arc.
   D. Aluminum melts before it changes color.

4. _____

5. Answer the following two questions for the welding of aluminum using GTAW.
   A. What type of current can be used?

   _____

   B. What type of tungsten electrode should be used with alternating current?

   _____
   _____

   C. What type of electrode should be used with direct current?

   _____

6. For each of the following, select the aluminum filler alloy best suited for the alloy combination and criteria. Refer to Figure 21-2 in the text.
   A. For welding 1100 aluminum to itself, which filler alloy provides the best strength?
   B. For welding 2219 aluminum to itself, which filler alloy gives the best results overall?
   C. For welding 6061 aluminum to itself, which filler alloy is easiest to weld with?
   D. For welding 6061 aluminum to 5052, which filler alloy has the best ductility and ease of welding?
   E. For welding 5052 aluminum to itself, which filler alloy has the best strength, ductility, ease of welding, and color match after anodizing?

6. A. _____
   B. _____
   C. _____
   D. _____
   E. _____

7. Which metal transfer methods are recommended for gas metal arc welding of aluminum? Why?

   _____
   _____
   _____

8. *True or False?* The preheat temperature for welding of thick section aluminum castings should be 400°F–500°F (204°C–260°C).

8. _____

9. List the two most widely used and recommended processes for welding aluminum.

9. _____

10. The metals that are alloyed to form bronze are _____ and _____.

10. _____

11. *True or False?* Only oxygen-bearing copper can be fusion-welded easily.

11. _____

Lesson 21 Nonferrous Welding Applications

Name _____

12. Answer the following two questions about welding copper using GTAW.
    A. What type of current should be used?
    _____
    B. What type of tungsten electrode should be used?
    _____

13. *True or False?* Because copper has a high thermal conductivity, a preheat of 400°F–500°F (200°C–260°C) is often used.   13. _____

14. A(n) _____ type electrode is used for arc welding brass.   14. _____

15. Why are oxyfuel gas processes *not* recommended for welding bronze?
    _____
    _____

16. A(n) _____ must be worn whenever you enter a gas-filled area.
    _____

17. Label the parts of the titanium welding set-up shown.
    A. _____
    B. _____
    C. _____
    D. _____
    E. _____
    F. _____

18. What is the biggest problem with welding titanium and its alloys?
    _____
    _____

19. *True or False?* To increase the welding gas temperature for welding plastic, the gas flow is increased.   19. _____

20. How is heat for plastic welding supplied?
    _____
    _____

Job 21-1

Groove Welds and Fillet Welds on Aluminum Using GTAW

Name _____ Date _____

Class _____ Instructor _____

> **Learning Objective**
>
> In this job, you will make groove welds and fillet welds on aluminum using the GTAW process. You will create the welded assembly shown in the figure at the end of this job assignment.

1. Obtain 10 pieces of aluminum 6061, 5052, or 1100. Each piece should measure 1/8″ × 2″ × 6″ (3.2mm × 50mm × 150mm).
2. What filler metal provides the best ease of welding for the base metal you will be welding? Refer to Figure 21-2 in the text. 2. _____
3. What diameter electrode is recommended for 1/8″ (3.2mm) thick aluminum? Refer to Figure 8-21 in the text. 3. _____
4. Obtain six 36″ (910mm) lengths of filler metal, 3/32″ (2.4mm) diameter. The type of filler metal is given in Question 2. Also, obtain one pure tungsten electrode. The diameter is given in Question 3. Install the electrode in the torch.
5. Clean the base metal by dipping the aluminum in a cleaning solution or by cleaning the areas to be welded with a stainless steel brush.
6. Set the machine for AC welding. Adjust the current and the argon flow as described in Figure 8-21 in the text.
7. Complete all welds shown in the following drawing. Follow the welding order outlined in Steps 8 through 20.

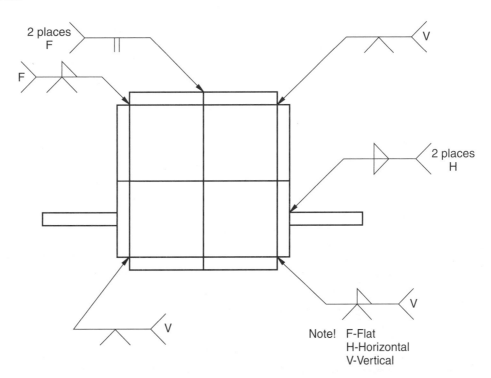

Copyright by The Goodheart-Willcox Co., Inc.

8. Align two pieces of aluminum to form a 6″ (150mm) square-groove butt joint. Tack weld in the middle and on each end. Use a root opening of 1/16″ (1.6mm).
9. Weld the entire length of the butt joint.
10. Examine the butt weld and discuss the results with your instructor.
11. Use six pieces of aluminum to make three additional butt joints as described in Step 8. Weld the entire length of each joint.
12. Align one piece of aluminum on one of the butt-welded plates to form a T-joint. Tack weld in the middle and on each end. Place the T-joint in the horizontal position and make a fillet weld along one side of the T-joint. Rotate the weldment and weld the other side of the T-joint in the horizontal position.
13. Examine the fillet welds and discuss the results with your instructor.
14. Tack weld and weld the final piece of aluminum on a different butt-welded plate to form a T-joint. Complete each fillet weld in the horizontal position.
15. Align two of the butt-welded plate assemblies to form a corner joint. Refer to the figure in Step 7. Pay attention to which pieces are being aligned. Tack weld in three places, then weld the outside corner joint in the flat welding position.
16. Discuss the results of Step 15 with your instructor.
17. Weld a fillet weld on the inside of this corner joint in the flat welding position.
18. Align the other two butt-welded plates to form a corner joint. Weld the outside corner weld first, then the inside corner weld. Make both welds in the vertical position.
19. Align the two corner weld assemblies to form an enclosed box. Refer to the figure shown in Step 7. Tack weld the remaining corners of the box.
20. Weld the outside corner joints in the vertical welding position. See Heading 8.10 and Figures 8-57, 8-58, and 8-59 in the text for information on gas tungsten arc welding in the vertical welding position.

Inspection

Each weld should have beads with even widths and evenly spaced ripples. There should be no overlapping or undercutting. All butt welds must have 100% penetration. The final assembly should match the figure shown in Step 7.

Instructor's initials: _____

Lesson 22

Pipe and Tube Welding

Name _____ Date _____
Class _____ Instructor _____

Learning Objective
You will be able to describe the principles and terms used in pipe and tube welding. You will be able to prepare, set up, tack weld, and weld pipe and tube.

Instructions
Carefully read Chapter 22 of the text and study the figures in the chapter. Then answer the following questions.

1. *True or False?* If a pipe is said to be 4" in diameter, the outside of the pipe is greater than 4".

    1. _____

2. What is the wall thickness of a schedule 40, 8" (200mm) pipe?

    2. _____

3. Which of the following pipe materials has good high-temperature strength, requires special welding procedures, and requires annealing and postweld heat treating?
    A. Carbon steel pipe
    B. High yield strength pipe
    C. Chrome-molybdenum pipe
    D. Plastic pipe

    3. _____

4. List two schedules of pipe that are considered standard pipe sizes.

    4. _____
    _____

5. *True or False?* Tubing is measured and ordered by its inside diameter (ID).

    5. _____

Copyright by The Goodheart-Willcox Co., Inc.   Modern Welding Lab Workbook   313

6. Identify the parts of a groove joint that are indicated on the following drawing. Refer to Figure 3-3 in the text.

   A. _____
   B. _____
   C. _____
   D. _____

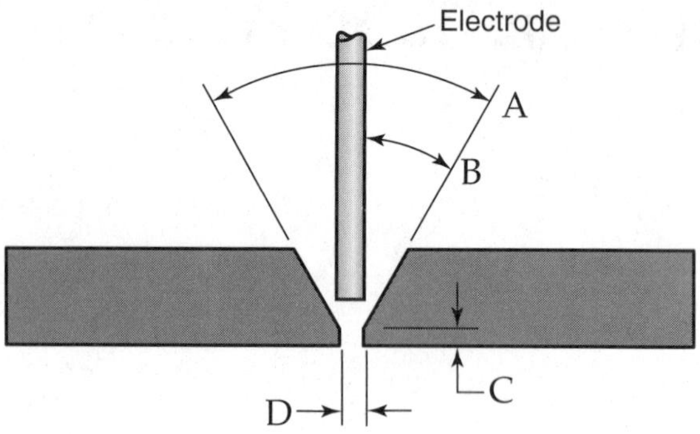

7. List three ways a pipe can be prepared (beveled) for welding.

   _____

   _____

   _____

8. What are the two directions in which a vertical butt joint can be welded?

   8. _____

9. Is penetration greater when welding uphill or downhill with SMAW?

   9. _____

10. What travel angle is recommended for welding a pipe in the 5G position?

    10. _____

11. *True or False?* The root pass in a pipe weld is usually made with an E6010 or E7018 electrode, or with the GTAW process.

    11. _____

12. Label the different passes in the pipe weld shown.

    A. _____
    B. _____
    C. _____
    D. _____

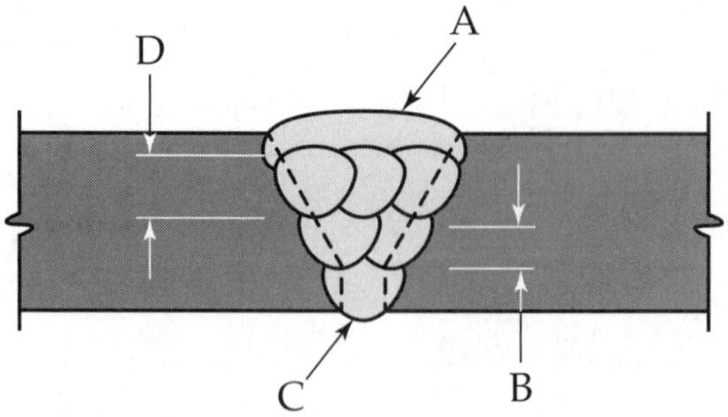

13. Other than E6010 and E7010, name an SMAW electrode commonly used for mild steel pipe welding.

    13. _____

Name _____

14. When an 8″ (200mm) diameter pipe with a wall thickness of 3/16″ (5mm) is welded with an E6010 electrode, what electrode size and current should be used? Refer to Heading 6.6.1 and Figure 6-17 in the text.

    Electrode size: _____

    Current: _____

15. When an E6010 electrode is used to weld a 10″ (250mm) schedule 10 pipe, what electrode size and current should be used? Refer to Heading 6.6.1 and Figures 6-17 and 22-3 in the text.

    Electrode size: _____

    Current: _____

16. When an E7018 electrode is used to weld a 10″ (250mm) schedule 10 pipe, what electrode size and current should be used? Refer to Heading 6.6.1 and Figures 6-18 and 22-3 in the text.

    Electrode size: _____

    Current: _____

17. What size welding tip orifice should be used to oxyacetylene weld a 2″ (50mm) diameter mild steel tube with a wall thickness of 1/16″ (1.6mm). Refer to Heading 12.2.3 and Figure 12-13 in the text.

    17. _____

18. What does the term "feathering" mean in pipe welding?
    A. Reducing the weld current at the end of the joint.
    B. Properly tying the toe of the weld into the pipe.
    C. Grinding a tack weld about 1/4″ (6.3mm).
    D. Buttoning the shirt collar before doing overhead welding.

    18. _____

19. What part of the GTAW torch contacts the pipe when "walking the cup"?

    19. _____

20. *True or False?* More than one welder can weld a large-diameter pipe at the same time.

    20. _____

21. The hot pass in pipe welds should be welded within ____ minutes after the root pass is completed.

    21. _____

22. List two tests used to determine if there is a leak in a pipe or tube.

    _____

    _____

23. *True or False?* Slag from the FCAW process is easier to remove than slag from the SMAW process.

    23. _____

24. Pipes or tubes to be welded must be absolutely free of ____ liquids or gases in order to prevent an explosion.

    24. _____

25. Label each of the positions shown with the correct AWS number and letter designation. Refer to Figure 22-21 in the text.

    A. _____
    B. _____
    C. _____
    D. _____

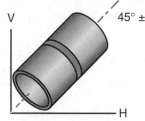

A—Multiple (not rotated)

B—Horizontal

C—Rotated flat

D—Multiple (not rotated)

Job 22-1

Welding Mild Steel Pipe in the 1G Position Using SMAW

Name _____   Date _____

Class _____   Instructor _____

> **Learning Objective**
>
> • In this job, you will weld a pipe joint in the 1G position using the SMAW process.

1. Obtain four sections of 4″ (100mm) diameter mild steel pipe about 3″ (75mm) long and about schedule 40. Other diameter pipe can be used. The pipe should be prepared for welding. Also obtain a number of 1/8″ (3.2mm) diameter E6010 electrodes.

2. What is the wall thickness of a 4″ (100mm) diameter schedule 40 pipe?   2. _____

3. If the pipe is not prepared for welding, the edges should be beveled as shown in Figure 22-8 in the text.

4. What amperage range should be used for a 1/8″ (3.2mm) diameter E6010 electrode? See Figure 6-17.   4. _____

5. Align two sections of pipe to form a V-groove butt joint. Leave a 1/16″ (1.6mm) root opening as shown in the following drawing.

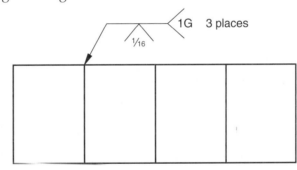

6. Tack weld the pipe in at least three places. Tack welds should be made in the flat position. Feather the tack welds.

7. Place the tack welded pipes into a weld positioner or use clamps to hold the pipes in the horizontal welding position. This is the 1G or 5G position.

8. For the purpose of describing pipe or tube welding, the circumference of the pipe or tube is divided like a clock face. The 12 o'clock position is at the top of the pipe or tube and 6 o'clock is at the bottom. Begin welding in the 3 o'clock position and weld the root pass to the 12 o'clock position. Refer to Figures 22-24 and 22-25 in the text for work and travel angles of the electrode.

9. After welding from the 3 o'clock position to the 12 o'clock position, stop and rotate the pipe 90°. The section welded is now in the 3 to 6 o'clock position.

Modern Welding Lab Workbook

10. Begin welding in the 3 o'clock position where the last weld ended, and weld to the 12 o'clock position. Be careful to properly fuse the new weld bead with the previous bead. Refer to Heading 6.6.4.

11. Continue rotating the pipe and welding from the 3 o'clock to the 12 o'clock position until the pipe is completely welded.

12. Depending on the wall thickness of the pipe you are welding, it may take more than one pass to complete the pipe weld. If a second weld pass is required, complete the root pass first. Then begin the second pass and continue to rotate and weld the pipe as described in Steps 8 through 11.

13. Align another pipe section onto the two pieces already welded and tack weld it.

14. Following the procedure described previously, completely weld this section of pipe.

15. Weld the last section of pipe onto those already welded.

Inspection

Examine the weld bead. Each bead should be even in width and smooth. Each start and stop must have good fusion with the previous bead. There should be no undercutting or overlapping. Examine the inside of the pipe. Penetration should be complete, but not excessive. Your instructor may have you perform a face bend, root bend, and tensile test on coupons removed from your pipe weld.

Instructor's initials: _____

Job 22-2

Welding Mild Steel Pipe in the 2G and 5G Positions Using SMAW

Name _____ Date _____

Class _____ Instructor _____

Learning Objective

- In this job, you will weld a pipe joint in the 2G and 5G positions using SMAW.

1. Obtain four sections of 4″ (100mm) diameter mild steel pipe about 3″ (75mm) long and about schedule 40. Other diameter pipe can be used. The pipe should be prepared for welding. Also obtain a number of 1/8″ (3.2mm) diameter E6010 and E7018 electrodes.

2. What is the wall thickness of a 4″ (100mm) diameter schedule 40 pipe?   2. _____

3. If the pipe is not prepared for welding, the edges should be beveled as shown in Figure 22-8 in the text.

4. What amperage range should be used for a 1/8″ (3.2mm) diameter E6010 electrode and an E7018 electrode? See Figures 6-17 and 6-18.   4. _____

5. Align two sections of pipe to form a V-groove butt joint as shown in the following drawing. Leave a 1/16″ (1.6mm) root opening.

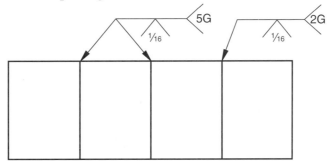

6. Tack weld the pipe in at least three places. Tack welds can be made in the flat position. Feather the tack welds.

7. Clamp the tack welded pipes into a weld positioner to hold the pipes in a horizontal orientation. There must be enough space to weld all around the pipe. This is the 5G position.

8. Begin welding in the 6 o'clock position, which is overhead welding. Weld from the 6 o'clock position to the 12 o'clock position, which will be uphill welding.

9. After reaching the 12 o'clock position, stop and restart at the 6 o'clock position. Weld uphill again to complete the pipe weld. *Do not rotate the pipe.*

10. If more than one pass is required to complete the pipe weld, weld the root pass completely. Remove all slag from the root bead. Change to an E7018 electrode and weld additional passes to complete the second pass or cover pass. Change the current setting on the welding machine when changing to an E7018 electrode.

11. Tack weld another pipe section onto the pieces already welded.

12. Follow the procedure described previously and completely weld this pipe joint.

13. Align another section of pipe to form a V-groove butt joint as described in Step 5 and tack weld the joint.

14. Place the pipes into a weld positioner vertically, so the joint is in the 2G position.

15. Weld the sections of pipe together. Use E6010 for the root pass and E7018 to complete the joint. Refer to Figure 6-49 in the text for the suggested electrode positions. Remember to point the electrode upward slightly to keep the weld pool from sagging.

16. Weld the last section of pipe onto those already welded. Weld it in the 2G position.

Inspection

Examine the weld beads. The beads should be even in width and smooth. Each start and stop must have good fusion with a previous bead. There should be no undercutting or overlapping. Examine the inside of the pipe. Penetration should be complete, but not excessive. Your instructor may have you perform a face bend, root bend, and tensile test on coupons removed from your pipe weld.

Instructor's initials: _____

Job 22-3

Welding Mild Steel Pipe in the 5G Position Using GTAW

Name _____   Date _____
Class _____   Instructor _____

> **Learning Objective**
> In this job, you will weld a pipe joint in the 5G position using the GTAW process.

1. Obtain four sections of 4″ (100mm) diameter mild steel pipe 3″ (75mm) long and about schedule 40. Other diameter pipe can be used. The pipe should be prepared for welding. Also obtain a 3/32″ (2.4mm) diameter 2% thoriated tungsten electrode or another type appropriate for the material you are welding. Obtain a number of pieces of 3/32″ (2.4mm) filler rod.

2. What is the wall thickness of a 4″ (100mm) diameter schedule 40 pipe?   2. _____

3. If the pipe is not prepared for welding, the edges should be beveled as shown in Figure 22-8 in the text.

4. What welding current is recommended for a 3/32″ (2.4mm) diameter electrode welding a 1/8″ thick butt joint in steel? Refer to Figure 8-20.   4. _____

5. What polarity current is recommended? Refer to Figure 8-20.   5. _____

6. What type shielding gas and flow rate are recommended for mild steel?   6. _____

7. Set the machine for the correct current and current type. (The electrode current range is given in Question 4 and the current type is given in Question 5.)

8. Align two sections of pipe to form a V-groove butt joint. Leave a 1/16″ (1.6mm) root gap as shown in the following drawing.

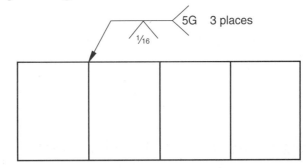

9. Tack weld the pipe in at least three places. Tack welds can be made in the flat position.

10. Clamp the tack welded pipes so they are oriented horizontally. The weld joint is in the vertical position. There must be enough room to weld all around the pipe.

11. For the purpose of describing pipe or tube welding, the circumference of the pipe or tube is divided like a clock face. The 12 o'clock position is at the top of the pipe or tube and 6 o'clock is at the bottom. Begin welding at the 5 o'clock position and weld uphill and over the top of the pipe to the 11 o'clock position. To complete the weld, begin at the 5 o'clock position and weld under the pipe and then weld uphill to the 11 o'clock position.

12. Depending on the thickness of the pipe you are welding, it may take more than one pass to complete the weld. Make additional passes as needed.

13. Align another pipe section onto the two pieces already welded and tack weld it.

14. Repeat the procedure described in Steps 8–12 to completely weld this section of pipe.

15. Weld the last section of pipe onto those already welded.

Inspection

Examine the weld beads. The beads should be even in width and smooth. Each start and stop must have good fusion with a previous bead. There should be no undercutting or overlapping. Examine the inside of the pipe. Penetration should be complete, but not excessive. Your instructor may have you perform a face bend, root bend, and tensile test on coupons removed from your pipe weld.

Instructor's initials: _____

Job 22-4

Welding Mild Steel Pipe in the 2G and 5G Positions Using GMAW

Name _____ Date _____
Class _____ Instructor _____

Learning Objective
- In this job, you will weld a pipe joint in the 2G and 5G positions using the GMAW process.

1. Obtain four sections of 4″ (100mm) diameter mild steel pipe about 3″ (75mm) long and about schedule 40. Other diameter pipe can be used. The pipe should be prepared for welding.

2. If the pipe is not prepared for welding, the edges should be beveled as shown in Figure 22-8 in the text.

3. Use a 0.035″ (0.9mm) diameter electrode wire and the short-circuiting transfer method.

4. What is the recommended amperage (amperes) and voltage (volts) for a 0.035″ (0.9mm) diameter electrode wire? Refer to Figure 9-13 in the text.

    4. _____ amperes
       _____ volts

5. What type current is recommended for the electrode you will be using?

    5. _____

6. What type shielding gas and flow rate is recommended for mild steel when the short-circuiting transfer method is used? The root pass will be on 1/16″–1/8″ (1.6mm–3.2mm) thick steel. Refer to Figure 9-34.

    6. _____

7. Set the machine for the correct current and current type. Refer to Questions 4 and 5.

8. Align two sections of pipe to form a V-groove butt joint. Leave a 1/16″ (1.6mm) root gap as shown in the following drawing.

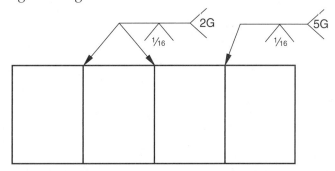

9. Tack weld the pipe in at least three places. Tack welding can be done in the flat position.

10. Clamp the tack welded pipes into a weld positioner so they are oriented vertically. The weld joint is in the horizontal welding position. There must be enough room to weld all around the pipe.

11. Begin welding at any point on the joint. Refer to Heading 9.10.3 in the text for horizontal welding techniques. Weld completely around the pipe. Whenever you stop and restart, make sure you fuse into the end of the prior weld bead.

12. Depending on the wall thickness of the pipe you are welding, it may take more than one pass to complete the pipe weld. If a second weld pass is required, complete the root pass first. Remove all slag. Then begin the second pass.

13. Align another pipe section onto the two pieces already welded and tack weld it.

14. Follow the procedure described previously and completely weld this section of pipe.

15. Remove the weldment from the fixture. Align and tack weld the final segment to the weldment. Tack welding can be done in the flat position.

16. Reposition the weldment in the fixture so the pipe is oriented horizontally.

17. Weld the last section of pipe onto those already welded.

Inspection

Examine the weld beads. The beads should be even in width and smooth. Each start and stop must have good fusion with a previous bead. There should be no undercutting or overlapping. Examine the inside of the pipe. Penetration should be complete, but not excessive. Your instructor may have you perform a face bend, root bend, and tensile test on coupons removed from your pipe weld.

Instructor's initials: _____

Job 22-5

Welding Mild Steel Pipe in the 6G Position Using GMAW

Name _____ Date _____

Class _____ Instructor _____

> **Learning Objective**
> - In this job, you will weld a pipe joint in the 6G position using the GMAW process.

> Note: The 6G position is the most difficult position in which to perform any pipe weld. Before attempting to weld in the 6G position, practice welding in the 2G and 5G positions until you can consistently produce high-quality welds. The 6G position is a combination of the 2G and 5G positions. Very few pipe welds on a construction site are made in this position. However, many pipe weld qualification tests use the 6G position. You should practice welding pipe in this position and become proficient at it before applying for a pipe welding job.

1. Obtain four sections of 4″ (100mm) diameter mild steel pipe 3″ (75mm) long and about schedule 40. Other diameter pipe can be used. The pipe should be prepared for welding.

2. If the pipe is not prepared for welding, the edges should be beveled as shown in Figure 22-8 in the text.

3. Refer to Headings 9.10, 9.11, and 9.12 in the text for welding techniques.

4. Determine the following values required to weld this joint:

   Type of electrode _____

   Metal transfer method _____

   Electrode diameter _____

   Amperage range _____

   Voltage range _____

   Suggested shielding gas _____

   Flow rate _____

   Type of DC current _____

5. Set the machine for the correct current and current type.

6. Align two sections of pipe to form a V-groove butt joint. Leave a 1/16″–3/32″ (1.6mm–2.4mm) root gap as shown in the following drawing.

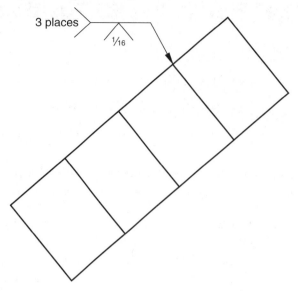

7. Tack weld the pipe in at least three places. Tack welds can be made in the flat welding position.

8. Clamp the tack welded pipes at a 45° angle in a weld positioner. The weld joint is also at a 45° angle. There must be enough room to weld all around the pipe.

9. Begin welding at the bottom of the pipe (the 6 o'clock position). Weld uphill to the top. As you move toward the top, the position and the angle of the electrode constantly changes. This is why the 6G position is so difficult. The electrode should be tipped 20° upward to counteract the sag of the molten metal. As you weld from the 8 o'clock to the 10 o'clock position, the electrode must also be tipped 20° upward to counteract the sag of the molten metal. When you reach the top of the pipe, stop. Restart at the bottom of the pipe and again weld uphill to the top. Whenever you stop, clean the bead. When you restart, be sure to fuse with the prior weld bead.

10. Depending on the wall thickness of the pipe you are welding, it may take more than one pass to complete the pipe weld. If a second weld pass is required, complete the root pass first. Wire brush the root pass to remove any glasslike coating or islands. Then begin the second pass.

11. Align another pipe section onto the two pieces already welded and tack weld it.

12. Repeat Steps 9 and 10 to completely weld this section of pipe.

13. Tack weld and weld the last section of pipe onto those already welded.

Inspection

Examine the weld beads. The beads should be even in width and smooth. Each start and stop must have a good fusion with a previous bead. There should be no undercutting or overlapping. Examine the inside of the pipe. Penetration should be complete, but not excessive. When done properly, there should be no difference between a pipe weld made in the 6G position and one made in any other position. Your instructor may have you perform a face bend, root bend, and tensile test on coupons removed from your pipe weld.

Instructor's initials: _____

Job 22-6

Welding Mild Steel Pipe in the 6G Position Using SMAW

Name _____ Date _____

Class _____ Instructor _____

> **Learning Objective**
> In this job, you will weld a pipe joint in the 6G position using the SMAW process.

> Note: The 6G position is the most difficult position in which to perform any pipe weld. Before attempting to weld in the 6G position, practice welding in the 2G and 5G positions until you can consistently produce high-quality welds. The 6G position is a combination of the 2G and 5G positions. Very few pipe welds on a construction site are made in this position. However, many pipe weld qualification tests use the 6G position. You should practice welding pipe in this position and become proficient at it before applying for a pipe welding job.

1. Obtain four sections of 6" (150mm) diameter mild steel pipe about 3" (75mm) long and about schedule 40. Other diameter pipe can be used. The pipe should be prepared for welding.

2. If the pipe is not prepared for welding, the edges should be beveled as shown in Figure 22-8 in the text.

3. Refer to Headings 6.9.4, 6.9.5, 6.9.6, 6.9.7 and 22.4 in the text for welding techniques.

4. Determine the following required to weld this joint:
   Type of electrode _____
   Electrode diameter _____
   Amperage range _____
   Type of dc current for E6010 and E7018 electrodes _____

5. Set the machine up for the correct current and current type.

Modern Welding Lab Workbook

6. Align two sections of pipe to form a V-groove butt joint. Leave a 1/8" (3.2mm) root gap as shown in the following drawing.

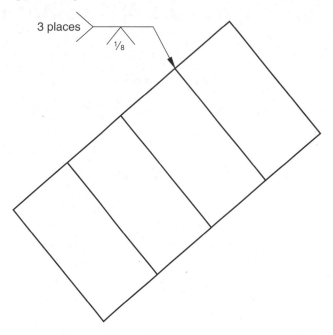

7. Tack weld the pipe in at least four places. Tack welds can be made in the flat position. Feather the tack welds.

8. Clamp the tack welded pipes at a 45° angle in a weld positioner. The weld joint is also at a 45° angle. There must be enough room to weld all around the pipe.

9. Begin welding at the bottom of the pipe (the 6 o'clock position). Weld uphill to the top. As you move toward the top, the position and the angle of the electrode constantly changes. This is why the 6G position is so difficult. When you reach the top of the pipe, stop. Restart at the bottom of the pipe and again weld uphill to the top. Whenever you stop and restart, make sure to remove all slag, especially from the toe of the weld where the weld meets the pipe. Slag can get trapped in this location.

10. Depending on the wall thickness of the pipe you are welding, more than one pass may be required to complete the pipe weld. If a second or additional passes are required, complete the root pass first. Remove all slag, and then begin the next pass.

11. Align another pipe section onto the two pieces already welded and tack weld it.

12. Repeat Steps 9–11 until all four pieces have been added to the weldment.

Inspection

Examine the weld beads. The beads should be even in width and smooth. Each start and stop must have a good fusion with a previous bead. There should be no undercutting or overlapping. Examine the inside of the pipe. Penetration should be complete, but not excessive. When done properly, there should be no difference between a pipe weld made in the 6G position and one made in any other position. Your instructor may have you perform a face bend, root bend, and tensile test on coupons removed from your pipe weld.

Instructor's initials: _____

Lesson 23A

Special Cutting Safety

Name _____ Date _____
Class _____ Instructor _____

> **Learning Objective**
> - You will be able to safely use special cutting equipment.
>
> **Instructions**
> *Carefully read Chapter 23 and pay attention to safety items. You may also need to refer to Chapter 11. Then answer the following questions.*

1. The oxygen cylinder valve is constructed using a back-seating valve to seal the stem from leakage. Therefore, when an oxygen cylinder is used, the valve must be _____.
   A. fully closed
   B. half-way open
   C. fully open
   D. whatever is required for the correct pressure

   1. _____

2. *True or False?* Cylinders must always be kept valve end up. When the cylinder is not in use, the cylinder valve should be closed whether the cylinder is full or empty.

   2. _____

3. Clothing that has been saturated with oxygen becomes highly flammable. It should be removed and not worn again for at least _____ minutes, or until no oxygen remains in it.

   3. _____

4. Oxygen regulators are closed when the regulator adjusting screw is turned all the way _____.

   4. _____

5. *True or False?* You should always wear safety (chipping) goggles when cleaning metals.

   5. _____

6. What number filter lens is required for air carbon arc cutting or gouging?

   6. _____

7. List the three elements needed for a fire.

Copyright by The Goodheart-Willcox Co., Inc.   Modern Welding Lab Workbook   **329**

8. *True or False?* More sparks are thrown back toward a welder in a cutting operation than in a piercing operation.

8. _____

9. List four pieces of protective equipment that must be worn for exothermic cutting operations.
   _____
   _____

10. *True or False?* To protect your eyes during flame- and arc-cutting, safety glasses or goggles plus a welding helmet or face shield may be necessary.

10. _____

11. *True or False?* High-pressure water will not cut any body parts.

11. _____

12. Two pieces of equipment that can be worn to prevent inhalation of contaminated air are an air-purifier and _____ breathing equipment.

12. _____

13. *True or False?* It is necessary to wear ear protection during exothermic and oxygen lance cutting.

13. _____

14. What is the function of a fire watch? _____
    _____

15. Which of the following should *not* be used to carry metal powder in a metal powder cutting operation?
    A. Nitrogen
    B. Air
    C. Oxygen
    D. All of the above.

15. _____

Lesson 23B

Special Cutting Processes

Name _____ Date _____

Class _____ Instructor _____

Learning Objective
You will be able to identify equipment and supplies used in special cutting and gouging processes and be able to describe the principles involved. You will also know how to choose and set the variables required to produce a quality cut.

Instructions
Carefully read Chapter 23 and study the figures in the chapter. You may also need to refer to Chapter 11. Then answer the following questions.

1. Which of the following is *not* required for air carbon arc cutting?
    A. Carbon or graphite electrodes
    B. Cylinder of oxygen
    C. Air carbon arc torch with air jet holes
    D. AC or DC arc welding machine

    1. _____

2. A constant _____ welding machine is required for manual air carbon arc cutting.

    2. _____

3. Name the parts in the air carbon arc cutting equipment shown in the following drawing.
    A. _____
    B. _____
    C. _____
    D. _____
    E. _____
    F. _____

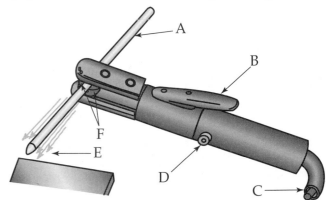

4. When cutting or gouging with CAC-A, the air jet holes should be _____ the electrode.

4. _____

5. Determine the amperage and travel speed required when a 3/8″ diameter electrode is used to make a 3/8″ deep gouge.

5. Amperage _____

   Travel Speed _____

6. What is the minimum and maximum current for a 3/8″ (9.5mm) carbon arc electrode used with DC?

6. _____

7. What travel angle is used for gouging with the air carbon arc process?

7. _____

8. When steel is cut using CAC-A, how far should the end of the electrode extend beyond the electrode holder?

8. _____

9. What temperature is achieved during the exothermic cutting process?

9. _____

10. The volume of oxygen required for exothermic cutting is _____ to _____ ft³/min (_____ to _____ l/min).

10. _____
    _____

11. What oxygen pressure and flow rate are required for a .675″ burning bar?
    _____

12. Describe the two main steps required to start an exothermic cut with a burning bar.
    _____
    _____
    _____

13. How is an exothermic rod ignited?
    _____

14. What travel angle is used to gouge with an exothermic rod?
    A. 90°
    B. 80°–85°
    C. 45°
    D. 20°–45°

14. _____

15. How does oxygen lance cutting differ from exothermic cutting?
    _____
    _____
    _____

16. *True or False?* Chromium oxide is introduced into the cutting flame during metal powder cutting.

16. _____

Lesson 23B   Special Cutting Processes

Name _____

17. What is the function of the chemical fluxes in chemical flux cutting?
    _____
    _____

18. What amperage and what type current is required to cut with a 5/32" (4.0mm) shielded metal arc cutting electrode?

    18. Amperage _____
        Current _____

19. Which of the following is *not* an advantage of laser beam cutting?
    A. Manual cutting is done at a high travel speed.
    B. A very narrow kerf is cut.
    C. The metal being cut is not part of the circuit.
    D. The laser beam can be deflected by mirrors.

    19. _____

20. Which of the following is not an orifice material used in water jet cutting?
    A. Diamond
    B. Ruby
    C. Marble
    D. Sapphire

    20. _____

Copyright by The Goodheart-Willcox Co., Inc.   Modern Welding Lab Workbook

Job 23B-1

Inspecting a CAC-A Welding Station

Name _____ Date _____

Class _____ Instructor _____

> **Learning Objective**
> - In this job, you will inspect an air carbon arc (CAC-A) welding station.

The equipment used in a CAC-A station includes the following:
- An AC or DC welding power source.
- An air compressor capable of delivering 80 to 100 psig (550 to 700 kPa) or a compressed air cylinder.
- A combination electrode lead and air hose.
- Workpiece lead.
- CAC-A electrode holder.
- Ventilation system.
- Booth or screening material.

1. Examine the combination electrode lead and air hose for cuts or worn areas.
2. Check that the compressor filter is clean. If a compressed air cylinder is used, make sure it is closed.
3. Check that the adjusting knob on the air pressure regulator is turned out and loose, so that the regulator is closed.
4. Check to ensure that the electrode and combination leads are tight on the power source.
5. Hang the CAC-A electrode holder on an insulated hook.
6. Turn on the ventilation system and check that it is working in your area.
7. Ensure that there are no holes in the work booth. If cutting is to be done in the open, place protective screening around the work area.
8. Note any discrepancies on the lines provided.

_____
_____
_____
_____

Instructor's initials: _____

Job 23B-2

Cutting and Piercing Using CAC-A

Name _____   Date _____

Class _____   Instructor _____

> **Learning Objective**
>
> • In this job, you will cut and pierce plain carbon steel using the air carbon arc (CAC-A) process.

1. Obtain a piece of mild carbon steel that measures 1/2″ × 3″ × 10″ (13mm × 75mm × 250mm).
2. Obtain a 1/4″ (6.4mm) air carbon arc electrode.
3. Set up the CAC-A station with the following variables:
   A. For a DCEP electrode, the amperage should be ____.
   B. For an AC electrode, the amperage should be ____.
   C. The air pressure fed to the CAC-A holder should be ____.
   D. The angle of the electrode to the base metal should be ____.

   3. A. _____
      B. _____
      C. _____
      D. _____

   ■ **Note:** Piercing with the CAC-A process is done in the same manner as is done using SMAC. Reread Heading 23.12 before proceeding.

4. Pierce four holes approximately 1/2″ (13mm) in diameter along a line 1 1/2″ (40mm) from the edge of your plate as shown in the following drawing.

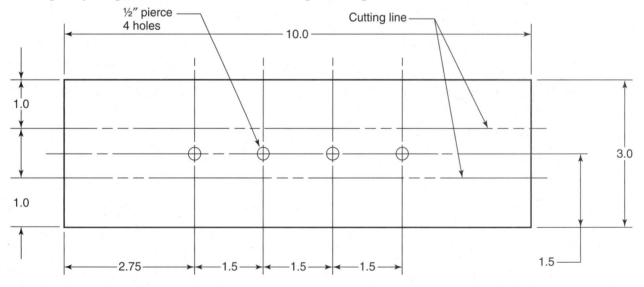

5. Cut the 3″ (75mm) piece of steel in two places as shown in the drawing.

> **Inspection**
>
> The edges of the holes and cut surfaces can be rough and ragged, but they should be as straight as possible. The diameter of the holes and the location of the cuts should be within the tolerance limits set by your instructor.
>
> Instructor's initials: _____

Job 23B-3

Removing a Weld or Weld Reinforcement Using CAC-A

Name _____   Date _____

Class _____   Instructor _____

> **Learning Objective**
> - In this job, you will use the air carbon arc (CAC-A) process to remove a weld or the weld reinforcement from a weld bead.

1. Obtain from the scrap container three previously welded pieces of mild steel that are 3/8" (9.5mm) or thicker. Ask your instructor to approve your choices.

2. To perform this CAC-A gouging operation, you will use a 5/16" (7.9mm) CAC-A electrode. Determine the following variables and set them on your cutting station:
   A. For DCEP, the amperage should be _____.
   B. For AC, the amperage should be _____.
   C. The recommended air pressure is _____ psig or _____ kPa.
   D. What travel angle is recommended for gouging?

   2. A. _____
      B. _____
      C. _____
      D. _____

3. Inspect the area for safety. Make sure there are no flammable items in the area.

4. Using the correct angle for gouging, remove the weld reinforcement and gouge into the weld about 1/8" (3mm) on one of the three welds. This technique is used to gouge out a defect found in a weld.

5. Check the appearance of the base metal surface to determine if any changes are required to the variables, your technique, or travel speed.

6. Remove the weld reinforcement from the other two welds.

> **Inspection**
> Inspect each gouged weld bead. The weld reinforcement should be removed. A gouge about 1/8" (3mm) deep should be present on each bead. The depth and width of each gouge should be fairly uniform. Rough ripples will be present, but the depth and width should not vary too much. Ripples will be smoother with a steady travel speed and arc length.
>
> Instructor's initials: _____

Lesson 24

Underwater Welding and Cutting

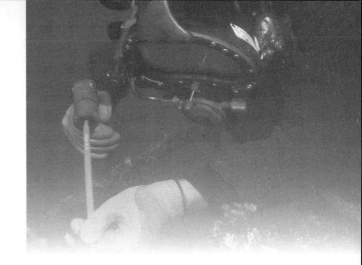

Name _____ Date _____

Class _____ Instructor _____

Learning Objective
- You will be able to demonstrate an understanding of underwater welding and cutting, the techniques used, terminology, and safety.

Instructions
Carefully read Chapter 24 of the text and study the figures in the chapter. Then answer the following questions.

1. What is the name of the AWS D3.6M specification?
   _____

2. Which weld class can be welded wet or dry, and are based on   2. _____
   AWS D1.1 Structural Welding Code?
   A. Class A
   B. Class B
   C. Class O
   D. None of the above.

3. Match each term with the appropriate statement or description.
   _____ GTAW
   _____ Background gas
   _____ Class O weld
   _____ Underwater SMAW electrode
   _____ Class B weld

   A. May have greater porosity and is used for less critical applications.
   B. Gas mixture used to force water out of an underwater habitat.
   C. Has waterproof coating over flux.
   D. Only performed in a dry habitat.
   E. Most demanding class of underwater welds.
   F. For constructing or repairing main structural members.

4. *True or False?* Dry underwater welding is referred to as   4. _____
   hyperbaric welding.

Copyright by The Goodheart-Willcox Co., Inc.   Modern Welding Lab Workbook   339

5. Which of the following are essential variables per AWS D3.6? Indicate all that are essential.

    5. _____

    A. Electrode manufacturer.
    B. Wet versus dry welding.
    C. Background gas.
    D. Electrode coating.
    E. Exposure time of electrode.
    F. Type of steel being welded.

6. What diameter SMAW electrodes should be used for wet welding?

    _____

7. Which of the following statements best describes *carbon equivalent*?

    7. _____

    A. A way to compare carbon content in various steels.
    B. A measurement of how much carbon is lost during underwater welding.
    C. A way to quantify the weldability of a steel.
    D. The amount of carbon in the flux of an SMAW electrode.

8. *True or False?* In wet welding, it is desirable to maintain a standard to long arc length.

    8. _____

9. What happens when a welder/diver calls to the surface and tells the supervisor or radio operator to "make the knife switch hot"?

    _____

    _____

    _____

10. A drag angle of _____ or less is used for SMAW wet welding.

    10. _____

11. There are two ways to end a wet SMAW weld bead. Which method does not require grinding the area prior to restriking the arc and continuing the weld bead?

    _____

    _____

12. Which of the following is not one of the three most commonly used processes in a dry habitat?

    12. _____

    A. GMAW
    B. SMAW
    C. GTAW
    D. SAW

13. Which of the following underwater habitats have access to the surface? Indicate all that apply.

    13. _____

    A. One-atmosphere pressure vessel.
    B. Ambient pressure chamber.
    C. Open bottom chamber.
    D. Cofferdam.

Lesson 24   Underwater Welding and Cutting

Name _____

14. Which of the following is not permitted in an underwater habitat?
    A. Mechanical cutting.
    B. Cutting with cutting torch.
    C. Carbon arc gouging.
    D. GTAW with argon/CO_2 shielding.

    14. _____

15. Wet carbon arc gouging is used to remove cracks from a weld. How much metal must a welder/diver grind off to remove the carbon and heat affected zone?

    _____

16. List the two categories of underwater cutting.

    _____

    _____

17. Mechanical cutting operations are commonly referred to as _____.

18. Which of the following statements is true?
    A. A guillotine saw can only be used to depths of 150′ (50m).
    B. Exothermic cutting is used to bevel a pipe prior to welding.
    C. A hydraulic milling machine can cut and bevel at the same time.
    D. A hydraulic lathe has one stationary ring and two or more rotating rings.

    18. _____

19. *True or False?* Exothermic cutting can cut through any steel, but not concrete.

    19. _____

20. *True or False?* In saturation diving, a welder/diver can stay underwater and under pressure for days, or even weeks, and then only requires one decompression cycle prior to returning to normal atmospheric pressure.

    20. _____

Lesson 25

Automatic and Robotic Welding

Name _____ Date _____
Class _____ Instructor _____

Learning Objective

- You will be able to describe the principles of automatic welding and robot systems.

Instructions

Carefully read Chapter 25 of the text and study the figures in the chapter. Then answer the following questions.

1. _____ is *not* an advantage of automatic welding.
   A. Increased productivity
   B. Increased job creation
   C. Improved quality
   D. Reduced cost

   1. _____

2. *True or False?* Flow switches are a form of feedback control.

   2. _____

3. A flow switch can be used to _____.
   A. sound an alarm
   B. shut off the machine
   C. light a warning light
   D. All of the above.

   3. _____

4. *True or False?* A relay uses a large voltage to control a smaller voltage.

   4. _____

5. *True or False?* All automatic welding operations depend on the use of solenoid-operated valves and relays.

   5. _____

6. Label the parts of the GMAW circuit shown in the following drawing.

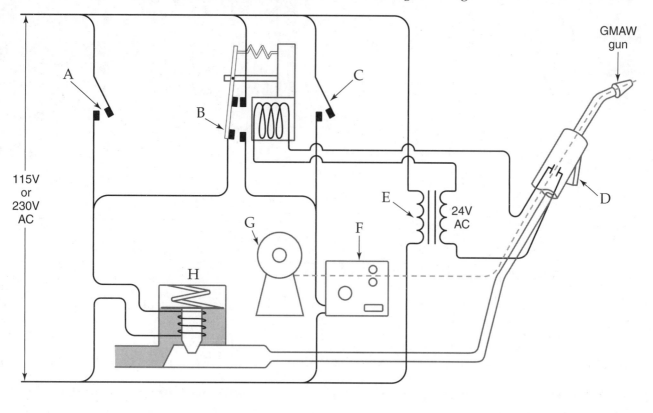

A. _____
B. _____
C. _____
D. _____

E. _____
F. _____
G. _____
H. _____

7. Describe what happens when the gun switch on a GMAW torch is pressed. _____
_____
_____
_____
_____

Lesson 25   Automatic and Robotic Welding

Name _____

8. Identify the parts of the robotic workcell shown in the following image.

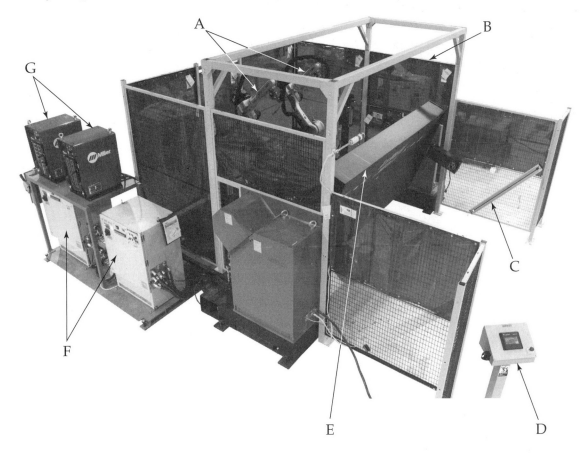

A. _____   E. _____

B. _____   F. _____

C. _____   G. _____

D. _____

9. Most welding robots have _____ axes.   9. _____
   A. four
   B. five
   C. six
   D. eight

10. Parts to be welded must be placed inside the robot's _____.   10. _____

11. The "brain" of a robotic operation is the _____.   11. _____

12. List two advantages of using a positioner.

_____

_____

13. A. What button is pressed to make the robot execute a program?
    B. What button is pressed if something goes wrong during the operation of the robot?

13. A. _____
    B. _____

14. *True or False?* When programming a robot, the teach pendant has control over each axis of the robot and positioner.

14. _____

15. *True or False?* Only an engineer or production supervisor should enter the working volume of a robot when the robot is in operation.

15. _____

Lesson 26
Metal Surfacing

Name _____ Date _____
Class _____ Instructor _____

Learning Objective
- You will be able to perform metal surfacing.

Instructions
Carefully read Chapter 26 of the text, and study the figures in the chapter. Then answer the following questions.

1. *True or False?* Hardfacing is the process of spraying surfacing material onto a base metal at extremely high temperatures.

   1. _____

2. Which of the following is *not* an advantage of metal surfacing?
   A. Certain dimensions can be maintained under adverse conditions.
   B. The service life of a part may be greatly increased.
   C. More expensive alloys are required to fabricate the part.
   D. Fewer replacement parts must be carried in stock.

   2. _____

3. *True or False?* Nickel-based alloy is one of the surfacing materials that can be used to resist factors that cause wear.

   3. _____

4. A part that will be subjected to chemical wear should *not* have a(n) ____ applied.
   A. nickel-base alloy
   B. aluminum-base alloy
   C. cobalt-base alloy
   D. tungsten carbide mixture

   4. _____

5. A(n) ____ flame is used when surfacing with an oxyfuel gas process.

   5. _____

6. Match each of the following processes with the corresponding advantage of that process.

   _____ Oxyfuel gas surfacing
   _____ Flux cored arc surfacing
   _____ Detonation flame spraying
   _____ Plasma arc spraying
   _____ Electric arc spray surfacing

   A. Molecules become ionized atoms that contain a great deal of energy.
   B. May be used for both thermal spraying and hardfacing applications.
   C. Involves maintaining an arc between two current-carrying wires.
   D. An ignition system sets off a charge several times per second.
   E. Hollow electrode wire contains powder.
   F. Electrode should be one size larger than is used for welding at similar currents.

7. Hardfacing with the oxyfuel gas process requires a tip that is ____ to ____ sizes larger than the tip used to weld with the same diameter rod.

   7. _____

8. *True or False?* An oxyfuel gas torch can be used for flame spraying.

   8. _____

9. For surfacing with the shielded metal arc process, a(n) ____ arc length should be maintained.

   9. _____

10. What is the recommended height and width of a deposit? How much should a second bead overlap the first?

    _____

    _____

    _____

11. Label the parts of the wire feed flame spraying torch shown in the following drawing.

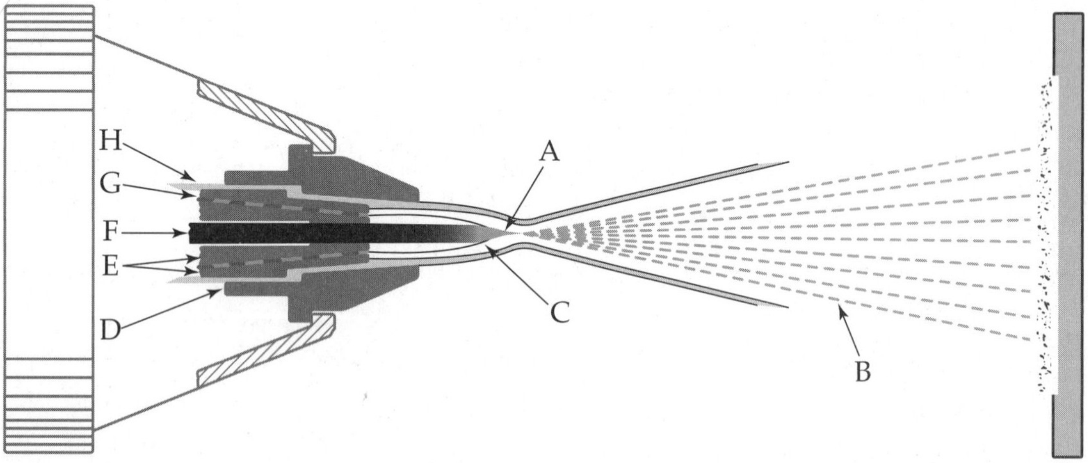

    A. _____
    B. _____
    C. _____
    D. _____
    E. _____
    F. _____
    G. _____
    H. _____

Name _____

12. Material to be flame sprayed is sold in different forms. Which of the following is *not* one of the available forms?
    A. Liquid
    B. Powder
    C. Wire
    D. Powder placed inside a hollow metal wire or plastic tube

    12. _____

13. Why is oxygen never used as a carrier gas in flame spraying?
    _____
    _____

14. *True or False?* Ceramic material can be flame-sprayed.

    14. _____

15. The tool in the following illustration is referred to as a wire-feed _____ torch.

    15. _____

16. *True or False?* Material is flame-sprayed in layers 0.001″ to 0.002″ (0.025mm to 0.050mm) thick.

    16. _____

17. Electric arc spraying uses _____ current carrying wires that are automatically fed to the arc position in the gun. Arc spraying guns can spray up to _____ lbs of wire per hour.

    17. _____
        _____

18. The plasma gas used in plasma arc spraying usually consists of 90%–95% _____ and 5%–10% _____.

    18. _____
        _____

19. List three types of surface preparation used to prepare a metal for flame spraying.
    _____
    _____

20. Identify some of the methods used to test and inspect surfacing materials once they have been applied to the base metal.

_____
_____
_____
_____
_____
_____
_____
_____

Job 26-1

Hardfacing a Steel Plate Using the Oxyfuel Gas Process

Name _____ Date _____
Class _____ Instructor _____

> **Learning Objective**
> • In this job, you will demonstrate your ability to hardface a steel plate using an oxyfuel gas torch.

1. Obtain one piece of mild steel plate 3/16" × 6" × 6" (4.8mm × 150mm × 150mm). Other thickness can be used.

   What type of flame should be used? _____

   What size welding tip should be used for welding 3/16" (4.8mm) plate? Refer to Figure 12-13 in the text.
   _____

   What size tip should be used to hardface? _____

   What diameter welding rod should be used to hardface? Refer to Figure 12-13 in the text. _____

2. Obtain six hardfacing rods.

3. The following drawing shows how the plate will be surfaced.

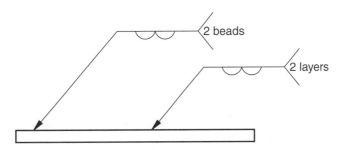

4. Light the torch and adjust it to the flame type entered as an answer in Step 1.

5. Heat one edge of the metal with the torch to 2200°F (1200°C). Steel is a bright red-orange color at this temperature. Do not melt the base metal.

6. After the metal has been heated, touch the hardfacing rod to the steel.

7. Run a bead down one side of the plate as if you were welding, but do not melt the base metal. Keep the metal at the correct temperature. The base metal, not the torch, should melt the rod. The bead can be run using either the forehand or backhand method. Refer to Figures 26-9 through 26-12 in the text.

8. After completing one bead, move over 3/4" (19mm) and run a second bead. The toes of the two beads should not touch.
   In an actual surfacing operation, how much should one bead overlap another?
   _____

9. Reposition the torch and rod and run a bead down the center of the plate.

10. Run a second bead so that it overlaps the first bead by the amount answered in Step 8.

11. Run a third and a fourth bead on the steel plate so each overlaps the previous by the correct amount.

12. Lay another bead that is centered between the two beads created in Steps 9 and 10. This way, the hardfacing material can be built up. Refer to Figure 26-6 in the text.

13. Lay two more beads between the beads made in Step 11. The buildup should be approximately 1/4″ (6.3mm).

Inspection

Examine the beads you have laid. Each bead should be even in width and uniform in height. There should be no visible cracks or chips in the beads. Any start and stop should have good fusion into the previous bead.

Instructor's initials: _____

Lesson 27

Production of Metals

Name _____ Date _____

Class _____ Instructor _____

Learning Objective
- You will be able to describe the production of many metals used in industry.

Instructions
Carefully read Chapter 27 of the text and study the figures in the chapter. Then answer the following questions.

1. List the two metals most commonly used in industry.
   _____

2. Steel that consists of just iron and carbon is referred to as ____.   2. _____

3. *True or False?* The atmosphere inside a steelmaking furnace is an oxidizing atmosphere.   3. _____

4. In order to make steel, the ____ in pig iron must be removed.   4. _____

5. Which of the following is *not* required to refine iron ore in a blast furnace?   5. _____
   A. limestone
   B. coke
   C. air
   D. sulfur

6. List the five major functions of a blast furnace.
   _____
   _____
   _____

7. *True or False?* In a blast furnace, the melting temperature of iron becomes lower as it combines with the excess carbon from coke.   7. _____

Copyright by The Goodheart-Willcox Co., Inc.   Modern Welding Lab Workbook   353

8. A(n) _____ is used to blow oxygen into the furnace during the basic oxygen furnace process.

8. _____

9. *True or False?* In an electric furnace, an arc is struck between flux cored electrodes and the metal in the furnace.

9. _____

10. *True or False?* The firebricks in an open hearth furnace are used to preheat the incoming fuel gases.

10. _____

11. A(n) _____ furnace uses the consumable electrode process to produce highly pure metal.

11. _____

12. To control temperature of the metal in an induction furnace, the _____ and the _____ of the alternating current must be controlled.

12. _____
   _____

13. Absorbed gases in steel cause porosity and inclusions. A(n) _____ furnace removes these gases.
    A. basic oxygen
    B. electric
    C. induction
    D. vacuum

13. _____

14. In the continuous casting process for the manufacture of steel, liquid steel from the furnace is moved in a ladle and poured into a reservoir called a(n) _____.

14. _____

15. Label the parts of the continuous casting process shown below.
    A. _____
    B. _____
    C. _____
    D. _____
    E. _____
    F. _____
    G. _____
    H. _____

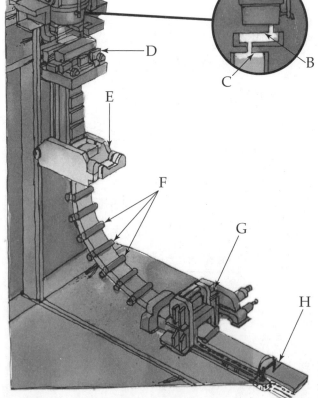

354   Modern Welding Lab Workbook

Name _____

16. Which of the following statements is *not* true? 16. _____
    A. Cast iron is usually made in a cupola furnace or an electric induction furnace.
    B. Gray cast iron is made by cooling an iron casting quickly.
    C. White cast iron is difficult to machine.
    D. The graphite in gray cast iron causes its gray appearance.

17. Copper is first refined in a(n) _____ furnace. It is further 17. _____
    refined using a(n) _____ process.

18. Brass and bronze are produced in a(n) _____. 18. _____
    A. induction furnace
    B. blast furnace
    C. vacuum arc furnace
    D. clay crucible

19. Aluminum is produced using an electrolysis process called 19. _____
    the _____ process.

20. Label the parts of the electrolytic cell shown in the following image.
    A. _____
    B. _____
    C. _____
    D. _____
    E. _____

21. List the five methods used to form metals into a finished or semifinished shape.
    _____
    _____
    _____

Lesson 28

Metal Properties and Identification

Name _____ Date _____

Class _____ Instructor _____

> **Learning Objective**
> You will be able to describe the properties of different metals and identify them.
>
> **Instructions**
> *Carefully read Chapter 28 of the text and study the figures in the chapter. Then answer the following questions.*

1. Match the following carbon contents with the correct designation for that carbon range.

   _____ 0.30% to 0.55% carbon
   _____ 0.003% carbon
   _____ less than 0.030% carbon
   _____ 1.8% to 4.0% carbon
   _____ 0.55% to 0.80% carbon

   A. Very high-carbon steel
   B. Low-carbon steel
   C. Cast iron
   D. Medium-carbon steel
   E. Wrought iron
   F. High-carbon steel

2. The physical properties of iron and steel are greatly affected by four factors. Name them. _____
   _____
   _____
   _____

3. Match each physical property listed with its correct description.

   _____ Compressive strength
   _____ Ductility
   _____ Toughness
   _____ Tensile strength
   _____ Brittleness
   _____ Elongation
   _____ Hardness

   A. How much a metal will stretch
   B. Resistance to being pulled apart
   C. Ease of fracturing
   D. Ability to be stretched
   E. Resistance to penetration
   F. Ability to prevent a crack from growing
   G. Ability to withstand a squeezing force

Copyright by The Goodheart-Willcox Co., Inc.   Modern Welding Lab Workbook   357

4. *True or False?* A metal that has over 5% elongation is considered to be ductile, and a metal that has less than 5% elongation is considered to be brittle.

4. _____

5. The following elements are often added to plain carbon steel to produce alloy steels. Match each element to its effect on steel.

   _____ Chromium
   _____ Nickel
   _____ Molybdenum
   _____ Vanadium
   _____ Tungsten
   _____ Manganese

   A. Increases toughness, increases strength at high temperatures
   B. Produces dense, fine grains
   C. Increases resistance to corrosion
   D. Increases strength and responsiveness to heat treatment
   E. Retards grain growth
   F. Increases strength, ductility, and toughness

6. Write a brief definition of the following terms.

   Solidus: _____
   _____

   Liquidus: _____
   _____

7. Label the parts of the iron-carbon diagram shown. Record your answers on the lines provided.

   ■ Note: The answers to F and G are *not* hypereutectoid and hypoeutectoid steels.

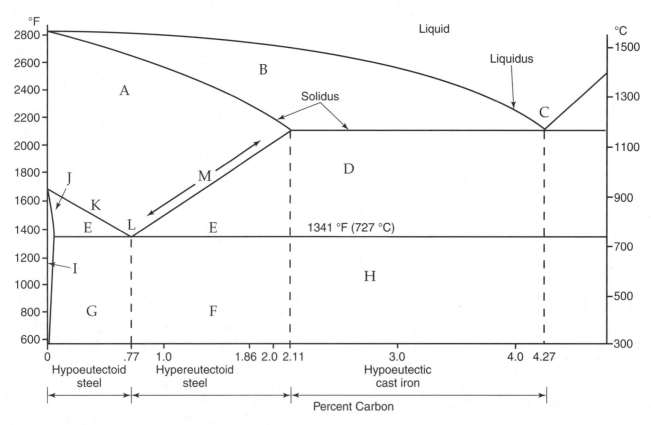

358   Modern Welding Lab Workbook

Lesson 28 Metal Properties and Identification

Name _____

A. _____   H. _____
B. _____   I. _____
C. _____   J. _____
D. _____   K. _____
E. _____   L. _____
F. _____   M. _____
G. _____

8. Which of the following is *not* true of ferrite?   8. _____
   A. Ferrite is ductile.
   B. Ferrite has a face-centered structure.
   C. Ferrite contains very small amounts of carbon.
   D. Ferrite is tough.

9. Pearlite is a combination of _____ and _____.   9. _____
   _____

10. *True or False?* Cementite contains 6.69% carbon and is hard   10. _____
    and brittle.

11. Pearlite always contains _____% carbon.   11. _____

12. At the eutectic point, solid austenite transforms into two new solids. List them. _____
    _____

13. The characteristics of the sparks in a spark test indicate the   13. _____
    _____ content of the steel or cast iron being tested.

14. List the four characteristics of the sparks in a spark test that indicate the nature of the steel being tested.
    _____
    _____

15. Which of the following results from an oxyacetylene torch test   15. _____
    would indicate that a metal is *not* suitable for welding?
    A. The weld pool is fluid and has good surface tension.
    B. The metal has an even, shiny appearance after
       solidification.
    C. The metal emits few sparks.
    D. The molten metal has a dull or colored surface.

16. List four tests, other than the spark test and oxyacetylene torch test, that are performed without
    specialized equipment and are used to help identify metals.
    _____
    _____
    _____
    _____

17. Nonferrous metals can be divided into what two general color classifications? _____
    _____

18. The surface of a fracture shows the grain structure of a metal. 18. _____
    Small grains indicate that the metal is _____.
    A. strong
    B. weak
    C. ductile
    D. unweldable

19. An AISI 5150 steel is a(n) _____ steel. 19. _____
    A. chromium
    B. carbon
    C. nickel
    D. tungsten

20. Which of the statements below is *not* true for nonferrous 20. _____
    metals?
    A. They have distinctive colors.
    B. They are magnetic.
    C. They are usually relatively soft.
    D. They do not spark when touched to a grinding wheel.

21. Which nonferrous metal listed in Figure 28-15 in the text has 21. _____
    the highest melting point?

22. List two metals that exhibit hot shortness.
    _____

23. A. Most copper is a(n) _____ color. 23. A. _____
    B. What is the melting temperature of copper? B. _____
    C. Is copper magnetic? C. _____

24. Which of the following is a characteristic of aluminum? 24. _____
    A. It has poor heat conductivity.
    B. It has good electrical conductivity.
    C. It changes color before it reaches its melting temperature.
    D. It oxidizes very slowly as it is heated.
    E. It melts at about 5000°F (2760°C).

25. Titanium retains its strength up to _____°F (_____°C). 25. _____
    _____

360   Modern Welding Lab Workbook

Job 28-1

Identification of Metals

Name _____ Date _____

Class _____ Instructor _____

Learning Objective
- In this job, you will perform several tests to identify various metals.

1. List eight tests that can be performed without special equipment to help identify metals. _____
   _____
   _____
   _____
   _____

2. Your instructor will supply a box containing ten or more different metals. These metals may include low-carbon steel, high-carbon steel, tool steel, stainless steel, cast iron, copper, aluminum, magnesium, lead, or other metals.

3. Take one of the metals from the box and perform as many tests as are needed to identify it. Perform the color and density tests to narrow down the possibilities. Perform the spark test to determine carbon and alloy content.

   ■ **Caution:** Always wear approved eye protection when grinding.

4. List the test(s) you performed to determine the metal type.
   _____
   _____

5. When you have determined the metal type, write it on the following lines. If it is steel, also list the approximate carbon content and any alloys it contains.
   _____
   _____

6. Select another piece of metal from the box and determine what type of metal it is. List the test(s) you performed on this metal. List the metal type.
   _____
   _____
   _____

7. Practice on as many different pieces as possible. When you feel confident of your abilities to identify different types of metal, contact your instructor.

8. Your instructor will select two different pieces of metal and ask you to identify them. You will be required to tell which tests you performed and why. List the tests you performed on each metal, your reasons for performing these particular tests, and the test results on the blank lines provided.

Metal #1

Test Used	Test Results
_____	_____
_____	_____
_____	_____
_____	_____

Reasons for choosing these tests: _____
_____
_____
_____

Metal #1
identified as

_____

Metal #2

Test Used	Test Results
_____	_____
_____	_____
_____	_____
_____	_____

Reasons for choosing these tests: _____
_____
_____
_____

Metal #2
identified as

_____

Instructor's initials: _____

Lesson 29
Heat Treatment of Metals

Name _____ Date _____
Class _____ Instructor _____

Learning Objective
You will be able to explain how various heat treatments are used to alter the properties of metals.

Instructions
Carefully read Chapter 29 of the text and study the figures in the chapter. Then answer the following questions.

1. Identify the three times that heat can be applied to a metal during the welding cycle.
   _____
   _____
   _____

2. The ____ is *not* an important factor to consider when heat treating.
   A. temperature to which a metal is heated
   B. the method used to heat the metal
   C. rate at which the metal is cooled
   D. material surrounding the metal when it is heated

   2. _____

3. There are many ways used to measure temperatures during heat treatment. Identify the following tool.

   _____

Copyright by The Goodheart-Willcox Co., Inc. Modern Welding Lab Workbook 363

4. List three methods (other than flame heating) that may be used to heat a part.

_____

_____

_____

5. *True or False?* The fastest method of cooling metal is immersion in a cold liquid.

5. _____

6. Heat loss takes place in a combination of three ways. Name all three.

_____

_____

_____

7. A steel that contains large amounts of _____ cannot be hardened because there is only a very small quantity of _____.

7. _____

_____

8. When steel is slow-cooled, _____ will form. Martensite is formed when steel is _____.

8. _____

_____

9. Describe the physical characteristics of each of the following microstructures.

Bainite: _____

_____

_____

Pearlite: _____

_____

_____

Martensite: _____

_____

_____

10. After steel is welded, there are two regions of grain sizes. The grain-_____ region is next to the weld metal. The grain-_____ region is next to the base metal.

10. _____

_____

11. The smallest grains are produced when steel is heated to the _____ critical temperature, and then cooled.

11. _____

12. *True or False?* Since grain growth is a function of time and temperature, a higher temperature and a longer time will produce a larger grain size.

12. _____

364 Modern Welding Lab Workbook

Name _____

13. Match each heat treatment listed with the statement that describes it.

   _____ Thermal stress relieving
   _____ Normalizing
   _____ Annealing
   _____ Spheroidizing
   _____ Quenching and tempering

   A. Heat to 1000°F–1200°F (537°C–749°C), which is below the A_1 critical temperature.
   B. Quench rapidly to form martensite, then reheat to improve toughness.
   C. The cementite forms into small, separate spheres.
   D. Furnace-cool steel to 50°F below the A_1 critical temperature.
   E. Cool in still air to room temperature.

14. List three reasons why steel would be annealed.

   _____
   _____
   _____

15. Why is martensite tempered in the quenching and tempering process?

   _____
   _____

16. *True or False?* The final microstructure of a spheroidized high-carbon steel has small, separate spheres of ferrite in cementite.

   16. _____

17. Which of the following heat treatments will produce the maximum hardness in steel?
   A. Heat to above the A_3 and slow cool.
   B. Thermal stress-relieve the steel.
   C. Heat to above the A_3 and rapidly quench.
   D. Leave the steel in the "as welded" condition.

   17. _____

18. A part to be casehardened is made out of _____ steel. The part is placed in a furnace that has a(n) _____ atmosphere.

   18. _____
       _____

19. *True or False?* White cast iron can be heat treated for 24 hours for each inch of thickness, and malleable cast iron will result.

   19. _____

20. What happens to the alloying elements in aluminum after a solution heat treatment?

   _____
   _____

Job 29-1

Heat Treating a Cold Chisel

Name _____   Date _____

Class _____   Instructor _____

Learning Objective

- In this job, you will practice heat treating a cold chisel. This process is described under section 29.9.1 of the text. Also read section 29.6 of the text.

■ **Safety Note:** Always handle the chisel with metal tongs to prevent burns. Only touch the chisel with your hands when this job is completed.

■ **Caution:** Wear approved eye protection throughout the heat treatment process.

1. Your instructor will provide a box with one or more cold chisels. Select one of the chisels. You will heat treat this chisel to obtain a hard cutting edge, hard and tough metal behind the cutting edge, and a hard and tough chisel body.

   What is the typical carbon content of a cold chisel? _____

   To what temperature should the chisel be heated initially? _____

2. If a heat treating furnace is available, turn the furnace on and set the temperature to 30°F (15°C) above the temperature a chisel should be heated initially (refer to answer in Step 1).

   ■ **Note:** The temperature always varies slightly in a furnace, so set the furnace slightly high to make sure the metal goes above its critical temperature.

3. The time that the chisel must be heated depends on the thickness. Place the chisel in the furnace until the entire chisel is at the correct temperature.

4. If a heat treating furnace is not available, use an oxyacetylene torch. Before heating, apply temperature indicating crayons or liquids to the chisel so you will know the temperature of the metal. You should apply two different temperature indicators—one for a temperature of 1350°F (730°C) and a second for a temperature of 1400°F (760°C).

5. Heat the chisel to the correct temperature. Keep the chisel at this temperature so that the entire tool is at the correct temperature. The amount of time depends on the thickness.

6. After heating the chisel for the correct amount of time, quickly insert the cutting edge of the chisel 1″ (25.4mm) into a bucket of cold water.

7. Leave the cutting edge in the water until the part in the water returns to a steel gray color. The rest of the chisel will still be a cherry red color.

8. Remove the chisel from the water.

9. Use a piece of emery cloth and polish the cutting edge of the chisel.

10. Heat from the body of the chisel will begin to reheat the cutting edge.

11. Watch the polished surface as it changes color. Colors will progress from the body of the chisel down toward the cutting edge. Wait until the cutting edge turns completely purple. This is about 600°F (315°C).

12. The body of the chisel will have cooled to below the critical temperature. Place the entire chisel in the cold water. Leave it in the water for 60 seconds.

13. Remove the chisel. You have just quenched and tempered the cutting edge of the chisel. The edge is very hard. The body is hard and tough due to the small grains in the metal.

Inspection

Your instructor will check the condition of the chisel by using it. Neither the cutting edge nor the body should break.

Instructor's initials: _____

Lesson 30

Inspecting and Testing Welds

Name _____ Date _____
Class _____ Instructor _____

Learning Objectives
- You will be able to explain the difference between a discontinuity and a defect.
- You will be able to use nondestructive inspection methods to determine if a weld contains any defects.
- You will be able to use destructive tests to evaluate the properties of a weld.

Instructions
Carefully read Chapter 30 of the text and study the figures in the chapter. Then answer the following questions.

1. *True or False?* All welds have discontinuities.   1. _____

2. Define the term "discontinuity."

   _____

   _____

   _____

3. Which of the following statements is *true*?   3. _____
   A. All discontinuities are defects.
   B. All defects are discontinuities.
   C. NDE stands for nondestructive evaluation.
   D. NDT is a nonstandard term for NDE.

4. Match each of the following abbreviations with the corresponding test name.
   _____ MT   A. Leak inspection
   _____ ET   B. Visual inspection
   _____ RT   C. Peel test
   _____ LT   D. Liquid penetrant test
   _____ UT   E. Ultrasonic test
   _____ PT   F. Magnetic particle test
   _____ VT   G. X-ray inspection
                      H. Eddy current inspection

Copyright by The Goodheart-Willcox Co., Inc.   Modern Welding Lab Workbook   369

5. Magnetic particle inspection can only be used to check for flaws at or near the ____, and can be used only on materials that can be ____.

5. _____
   _____

6. In ultrasonic testing, an electronic device called a(n) ____ is used to send ____ waves into the part to be tested.

6. _____
   _____

7. *True or False?* X-ray inspection can only be used to locate flaws at or near the surface of a metal.

7. _____

8. What values of a weld can be determined from a tensile test?
   _____
   _____

9. Label the parts of the hydraulic tensile test machine in the following illustration.
   A. _____
   B. _____
   C. _____
   D. _____
   E. _____

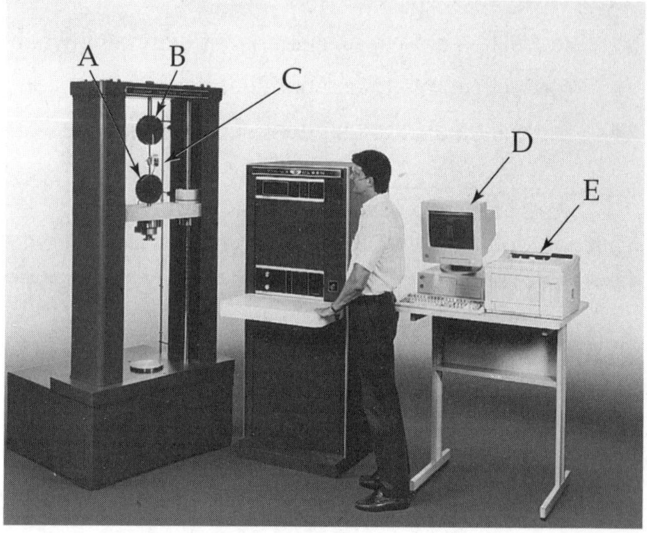

10. Match each of the following hardness tests with the corresponding description.
    _____ Rockwell B
    _____ Rockwell C
    _____ Brinell
    _____ Knoop

    A. The part must be sectioned with a highly polished surface.
    B. Uses a diamond ground to a 120° point.
    C. Uses a 1/16″ diameter ball indenter and a major load of 220 lbs.
    D. The load varies from 2.2 lbs to 220 lbs for a macro hardness test.
    E. Hardness number is calculated by dividing the applied load by the area of the indentation.

Name _____

11. Match each of the following test types or pieces of equipment to the corresponding destructive test category. (The test categories listed will be used more than once.)

    _____ Knoop diamond penetrator
    _____ Quantitative analysis
    _____ Izod method
    _____ Charpy method
    _____ Brinell
    _____ Qualitative analysis

    A. Chemical test
    B. Impact test
    C. Hardness test

12. *True or False?* Liquid or dye penetrant inspection can detect flaws both at the surface and below the surface.

    12. _____

13. Label the parts of the hardness tester shown below.

    A. _____
    B. _____
    C. _____
    D. _____
    E. _____
    F. _____

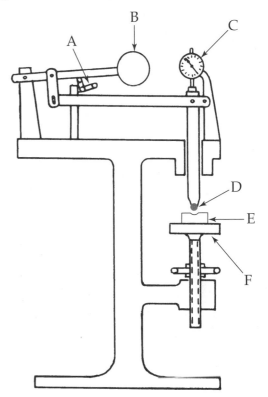

Job 30-1

Magnetic Particle Inspection

Name _____ Date _____

Class _____ Instructor _____

> **Learning Objective**
>
> • In this job, you will perform a magnetic particle inspection.

1. Obtain a piece of steel to be used for magnetic particle testing. The steel may be one of your welds from another job, or it may be a steel sample supplied by your instructor. Your instructor may have samples that are known to have cracks.

2. Examine the testing equipment.

   Does the equipment you will be using have a permanent magnet, an electromagnet, or test prods?
   _____

   Identify the parts of the magnetic particle test indicated in the following illustration.

   A. _____
   B. _____
   C. _____
   D. _____
   E. _____

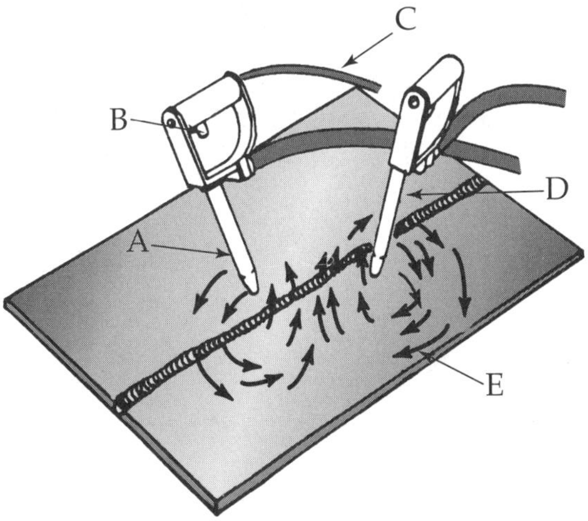

3. Clean the surface to be tested, using the cleaner supplied.

4. Lightly sprinkle or dust the magnetic particles on the area to be tested. Do not apply the magnetic material too heavily, or the particles will not be able to move.

   What happens to the sides of a small crack when a magnetic field is applied? _____
   _____
   _____

   What happens to the magnetic particles that are in the powder or liquid when a magnetic field is applied to the metal?
   _____
   _____

5. If a liquid is used, spray a light coating on the metal.

6. If using electrical prods or an electromagnet, turn on the power supply.

7. To test using electrical prods, place the prods in contact with the metal. Then, push the control switch. Keep the prods in contact with the metal.

8. To test using an electromagnet, place the yoke in contact with the metal and push the control switch. Keep the yoke in contact with the metal.

9. To test using a permanent magnet, just place the magnet on the metal. No switch is used.

10. With any of the methods used to perform the test, the results will be the same—the magnetic particles in the liquid or powder will gather around and point to any crack or flaw in the metal. Remember, even cracks or flaws that are below the surface (but very close to the surface) can be detected.

11. When testing with electrical prods or an electromagnet, it is often helpful to release the control switch so the magnetic field stops. Then, without moving the prods or yoke, push the control switch again. This will cause the magnetic particles to jump and move toward any crack or flaw.

12. Release the control switch and remove the prods or yoke.

13. Carefully examine the magnetic particles. They will gather around any flaw that can be detected. Remember, flaws that are parallel to the magnetic field will not be detected.

14. To finish the procedure, the part must be tested again. This time, move the test prods or yoke to a position that is 90° to the first test position.

15. Completely inspect the part you are testing. Each area should be tested twice—once with yoke or prods in one direction, the second time with the yoke or prods 90° to the first.

16. Mark any flaws you locate. You can mark them on the part, or make a sketch of the part and mark them on the sketch. Your instructor will ask you to show how you located the flaws. Once you have located all the flaws, contact your instructor.

Inspection

Your instructor will ask you to demonstrate your magnetic particle inspection technique by checking a certain area of the part. Your instructor will examine the area to verify that your results are correct.

Instructor's initials: _____

Job 30-2

Liquid Penetrant Inspection

Name _____ Date _____

Class _____ Instructor _____

Learning Objective
- In this job, you will perform a liquid penetrant inspection.

1. Obtain a part to be inspected. The part may be a weld from one of your other job assignments, or it may be a part supplied by your instructor. Your instructor may have parts that are known to have discontinuities (flaws) in them.

2. Clean the surface of the part with the cleaner provided.

3. Apply the penetrant to the surface. Wait 10 minutes so the penetrant can get into any cracks, pits, or voids.

4. Carefully read the instructions on the dye penetrant container. Use the recommended method to remove excess penetrant from the weldment. Describe the method used:

   _____
   _____
   _____

5. Apply the developer. Wait 10 minutes so that penetrant in any flaws will have time to be drawn out.

6. After waiting the recommended period, inspect the part. Most dyes will be visible in regular light. However, fluorescent penetrant must be inspected using an ultraviolet or "blacklight".

7. Record the locations of any flaws you discover. (Sketch the part and the location of flaws in the space provided.)

8. Clean the part after you have inspected it.
9. You may want to repeat the test on another piece of metal for practice.

Inspection

Your instructor may ask you to test another piece of material and locate all defects on it or to repeat the test on your practice piece. Perform the dye penetrant test as previously described. Carefully examine the part after the developer has been applied. Evaluate every indication and determine if it is a flaw. Many flaws, even very small ones, can be detected using penetrant inspection.

Instructor's initials: _____

Job 30-3

Guided Bend Tests

Name _____   Date _____

Class _____   Instructor _____

Learning Objective

- In this job, you will perform a guided bend test. Review section 30.2.1 in the text and study Figures 30-21 through 30-24.

1. Obtain one or more groove-welded plates. The plates may be any thickness, and should be at least 6″ (150mm) long.
2. If the plates are 3/4″ (19mm) thick or less, a face and root bend will be performed. If the plates are over 3/4″ (19mm) thick, side bends will be performed.
3. The radius to which the plates will be bent will be determined by your instructor.
4. If the plates are 3/4″ (19mm) thick or less, flame cut, plasma cut, or shear two pieces that are 1.5″ (40mm) wide, as shown in the following illustration.

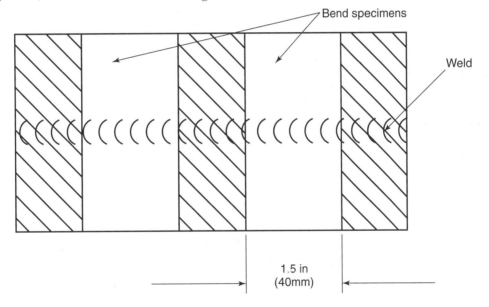

5. Use a power grinder or draw file to clean the cut edges and put a slight radius on the edge of the bend specimens. See the following sketch.

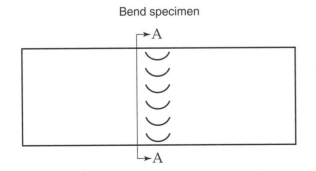
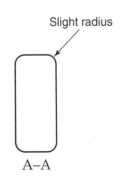

6. Your instructor will perform a face bend on one of your specimens. The instructor will perform a root bend on the other specimen.
7. If the plates are over 3/4″ (19mm) thick, flame cut four pieces that are 1/2″ (13mm) wide, as shown in the following illustration.

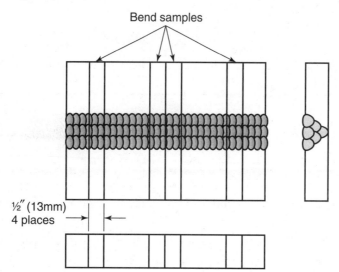

8. Use a power grinder or draw file to clean the cut edges and put a slight radius on the edge of the bend specimens. See the sketch in Step 5.
9. Your instructor will perform a face bend on one of your specimens. The instructor will perform a root bend on the other specimen. If your plate is over 3/4″ (19mm), all samples will be side bent.

   ■ **Caution:** Always wear approved eye protection when in an area where bend testing is being performed.

Inspection

Examine the bent specimens. There should not be any cracks or other defects over 1/8″ (3.2mm) long on the surface. For the purposes of this job, if a defect begins from the edge, it can be ignored if there is no evidence that it is due to poor welding techniques.

Instructor's initials: _____

Job 30-4

Tensile Testing

Name _____ Date _____

Class _____ Instructor _____

> **Learning Objective**
>
> In this job, you will prepare samples for tensile testing and will perform a tensile test on the samples. Review section 30.2.2 through 30.2.4 of the text and study Figures 30-27 through 30-29.

1. Obtain a groove-welded plate to be tensile tested. The sample can be one that you welded in another job assignment. The sample should be about 6″ × 6″ (150mm × 150mm).

2. You will need to prepare a reduced-section test specimen.

3. If the metal you will be testing is steel, you can flame cut or plasma cut the specimen. If the metal is not steel, you will need to plasma cut, saw, or shear the metal, and then have the reduced section machined.

4. Your instructor may have a template to follow for plasma or flame cutting. If a template is not available, mark your plate using chalk or soapstone, then cut the correct shape, as shown in the following illustration.

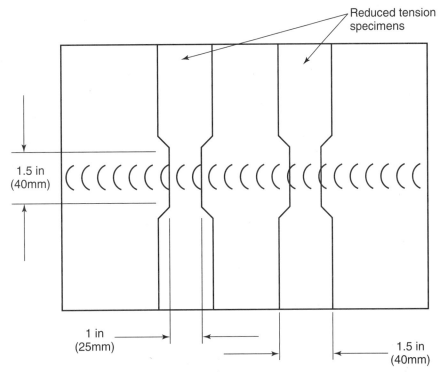

5. Grind the welds flush with the base metal on both sides of the plate.

6. Use a power grinder or draw file to round off the sharp edges on the plate. Grind or file any rough areas until they are smooth.

> **Caution:** Wear approved eye protection whenever you are in an area where grinding is being performed.

7. Measure the width and thickness of your samples. Enter the measured values in the spaces provided.

	Sample #1	Sample #2
Width of sample	_____	_____
Thickness of sample	_____	_____

8. Bring your prepared samples to your instructor for testing.

9. Your instructor will put your reduced section specimens in a tensile testing machine and apply a load to stretch the samples. Watch as the load (force) increases.

   ■ **Caution:** Wear eye protection while observing the test.

10. When your samples break, write down the load that broke each sample.

    Load to break sample #1 _____

    Load to break sample #2 _____

11. Examine the broken samples with your instructor. Where did the break occur? Was it in the weld metal, the heat-affected zone, or in the base metal?

    Sample #1 _____

    Sample #2 _____

12. Calculate the tensile strength of your samples. The formula to calculate the tensile strength is shown here. The formula is also shown in section 30.2.2 of the text with an example calculation.

$$\text{Tensile strength} = \frac{\text{tensile load}}{\text{area}} = \frac{\text{tensile load}}{\text{width} \times \text{thickness}}$$

■ **Note:** Width × thickness is the cross-sectional area of the sample.

Calculate the tensile strength of your samples and show your work in the space provided.

Tensile strength #1 _____

Tensile strength #2 _____

Inspection

The tensile strength of your samples should be equal to or exceed the tensile strength of the base metal. If the tensile strength of your samples is considerably less than the base metal, this indicates a problem in the welding technique. Examine the fractured edge to make sure there is no evidence of slag, inclusions, pits, or voids. If you have slag or other defects in your weld, but the tensile strength of your sample was equal to or exceeded the tensile strength of your base metal, your welds have good strength. However, you need to practice in order to eliminate these defects from your welding.

Instructor's initials: _____

Job 30-5

Hardness Testing

Name _____ Date _____

Class _____ Instructor _____

Learning Objective

● In this job, you will test the hardness of different metals.

1. Obtain four samples to use for hardness tests. The samples will be provided by your instructor. The samples may be low-carbon steel, high-carbon steel, copper, aluminum, or other metals. The samples may be the same metal, but have different heat treatments.

2. The Rockwell B scale is used on softer metals, like aluminum and copper. The Rockwell C scale is used on harder metals. The Brinell scale covers a large range from soft to hard metals.

3. First, the testing equipment must be checked to ensure that it is calibrated. Samples with a known hardness are supplied with the testing equipment. Test the hardness of a sample to verify that the equipment reads the hardness correctly. It is recommended that you confirm the calibration of the equipment every time you use it. You should also recalibrate the equipment for each metal type you will be testing. If the hardness tester does not read close to the value on the known hardness sample, inform your instructor. Your instructor will confirm your method and may adjust the equipment.

4. Test the hardness of the sample you have. You should take three hardness readings on each sample.

5. Record the hardness value for each reading you take in the following table. Average the three values and record the average. Look at the three hardness readings. The values should all be close together. If one reading is very high or very low, do not include it in your average.

Hardness Reading	Sample			
	#1	#2	#3	#4
#1				
#2				
#3				
Average				

Modern Welding Lab Workbook

Inspection

Your instructor will examine the three different hardness readings and the average. Your instructor will compare the average you obtained to the correct hardness value for that metal.

Instructor's initials: _____

Lesson 31

Procedure and Welder Qualifications

Name _____ Date _____
Class _____ Instructor _____

Learning Objective
- You will be able to use different codes, specifications, and standards to determine how welding should be performed. You will also be able to weld in accordance with a code or standard.

Instructions
Carefully read Chapter 31 of the text and study the figures in the chapter. Then answer the following questions.

1. Write the name of the agency, association, or society that corresponds with each of the following abbreviations.

    AWS _____

    FAA _____

    ASME _____

    ICC _____

    API _____

2. The transportation of all gas cylinders in the United States is regulated by the ____.
    A. National Bureau of Standards
    B. Association of American Railroads
    C. Interstate Commerce Commission
    D. American Welding Society

    2. _____

3. A code or specification must be followed when it is part of a(n) ____.

    3. _____

4. *True or False?* Every WPS must have a WPQR to document the quality of the welds produced using that procedure.

    4. _____

5. Which of the following variables is *not* an essential variable for an arc welding process?
   A. Base metal composition.
   B. Filler metal strength and composition.
   C. Welding position.
   D. Base metal thickness.
   E. Type of groove.
   F. Postheat temperature.

5. _____

6. *True or False?* Whenever there is a change in a nonessential variable, the WPS must be requalified.

6. _____

7. All welders must qualify for each welding _____ specification to which they will be required to weld.

7. _____

8. A welder performs a welding performance qualification test on 1″ (25mm) thick plate. Refer to Figure 31-3 in the text to answer Questions A, B, and C.

   A. On what range of base metal is the welder qualified to weld? _____
   B. How many tension specimens will be tested? _____
   C. How many side bends will be tested? _____

9. List three reasons why a welder may be required to requalify.
   _____
   _____
   _____

10. List the number and letter combination used to designate each of the welding positions shown in the following image.
    A. _____
    B. _____
    C. _____
    D. _____
    E. _____

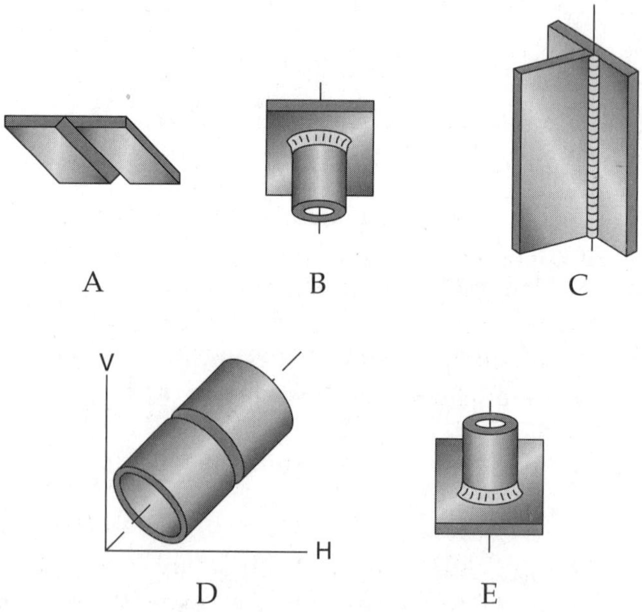

Job 31-1

Writing a Welding Procedure Specification and Qualifying the WPS

Name _____ Date _____

Class _____ Instructor _____

> **Learning Objective**
>
> In this job, you will establish and write a welding procedure specification (WPS) for welding a groove weld. You will attempt to qualify the WPS that you establish. You will weld a groove weld per your WPS. You will then remove samples and test them to determine if the WPS will make a quality weld. Welding will be done in the flat welding position. The material thickness will be 1/2" (13mm) or less.

Note to the Student

You will determine the values to be entered in the WPS. Each welder knows what electrode, what current, voltage, travel speed, and groove type will produce the best weld for him or her. From your knowledge of welding, determine for yourself what values will produce the best results. Enter these values on the WPS.

If you are not sure what values to enter, obtain two or more pieces of metal that your instructor assigns. Material thickness will be 1/2" (13mm) or less. Weld the pieces together using the process your instructor selects (SMAW, GMAW, GTAW, FCAW, OFW). Pay attention to the results of the parameters you use. After experimenting on practice pieces, enter the values that work best on the WPS. A sample WPS is provided at the end of this job.

1. You will be required to fill out all the information on the welding procedure specification form prior to doing any welding.

2. Enter the base metal type, base metal thickness, and process to be used as assigned by your instructor.

3. If you are using an oxyfuel gas process, enter the fuel and oxygen flow rates. If you are using a process that requires gas shielding, enter the type of gas shielding you will be using and the flow rate.

4. If you are using the oxyfuel gas or gas tungsten arc welding process, enter the filler rod type and filler rod diameter you will be using.

5. If you are using an arc welding process, enter the electrode type and electrode diameter. Note that a different type or diameter can be used for each pass. Enter the type and diameter you will use for each pass.

6. If you are using an arc process, enter the current, voltage, polarity, wire feed rate, and travel speed required for the process you are using.

7. Enter the type of groove you will be welding and the root gap, included angle, and root face.

8. If you will be using any preheat or postweld heat treatment, enter the appropriate temperatures.

9. In any spaces that are left blank, write "N/A" for *not applicable*.

10. Show your completed WPS to your instructor.

Copyright by The Goodheart-Willcox Co., Inc.   Modern Welding Lab Workbook

11. After reviewing your WPS, your instructor will give you two pieces of the metal you are to weld. The samples are to be 8" × 3" (200mm × 75mm) each.

12. Prepare the metal as stated on your WPS.

13. Obtain the type of electrode or filler metal you will use to weld.

14. Set up your welding station. Keep in mind all general shop safety rules and safety precautions required for the process you will be using.

15. Set the welding equipment to provide the parameters that you established on your WPS.

16. Tack weld the two pieces together.

17. When you are ready to weld, contact your instructor. Your instructor will review your machine setup and may watch as you weld to ensure that you are meeting the parameters listed on your WPS.

18. Cut two 1.5" (40mm) wide specimens from your completed weld for bend testing. Also cut one 1.5" (40mm) wide specimen for tensile testing. The location of each sample is shown in the following figure.

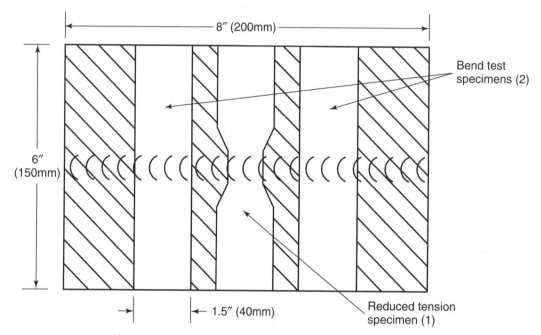

19. Prepare the two bend test samples.
    A. Use a power grinder or draw file to clean the cut edges and put a slight radius on the edge of the bend specimens. See the following sketch.
    B. Your instructor will perform a face bend on one of your specimens. The instructor will perform a root bend on the other specimen.

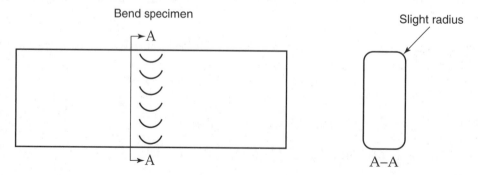

Job 31-1 Writing a Welding Procedure Specification and Qualifying the WPS

Name _____

20. Prepare the reduced-tension specimen. Refer to the following illustration.

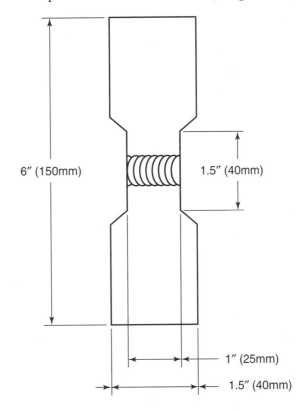

A. If the metal you will be testing is steel, you can flame cut the specimen. If the metal is not steel, you will need to saw or shear the metal, and then have the reduced section machined.
B. Your instructor may have a template to follow when flame cutting. If a template is not available, mark your plate using chalk or soapstone, then flame cut the correct shape.
C. Grind the welds flush with the base metal on both sides of the plate.
D. Use a power grinder or draw file to round off the sharp edges on the plate. Grind or file any rough areas until they are smooth.

21. Record the width and thickness of your reduced-tension specimen on the welding procedure qualification record (WPQR).

22. Bring the tensile and bend specimens to your instructor for testing. Wear approved eye protection in any area where testing is performed.

23. After the bend testing, have your instructor check "pass" or "fail" on the WPQR. After the tensile test is performed, calculate the tensile strength of the tensile specimen and record the value on the WPQR.

Inspection

Your instructor will review the results of the tensile test and bend tests to see if the weld made per your WPS is satisfactory. If the tensile and bend tests show a quality weld with good tensile strength and no welding defects, then the WPS is qualified.

Instructor's initials: _____

Welding Procedure Specification

Name _____ Date _____
Class _____ Instructor _____

Base metal type _____
Thickness _____
Process _____

Gas
Fuel gas flow _____
Oxygen flow _____
Shielding gas composition _____
Shielding gas flow _____

Electrode	Root Pass	2nd Pass	Fill Passes	Cover Pass
Electrode type	_____	_____	_____	_____
Electrode diameter	_____	_____	_____	_____
Filler metal type	_____	_____	_____	_____
Filler metal diameter	_____	_____	_____	_____
Welding Parameters				
Voltage	_____	_____	_____	_____
Current	_____	_____	_____	_____
Polarity	_____	_____	_____	_____
Travel speed	_____	_____	_____	_____
Wire feed rate	_____	_____	_____	_____
Contact tube to work distance*	_____	_____	_____	_____

Type of Groove
Type of groove _____
Root gap _____
Included angle _____
Root face _____

Heat Treatment
Preheat temperature _____
Postweld heat treatment temperature _____

*Used on automatic welding only

Welding Procedure Qualification Record (WPQR)

Tension Test

Width of specimen _____

Thickness of specimen _____

Load to break specimen _____

Tensile strength _____

Show your calculations. _____

Bend Test

	Pass	Fail
Face Bend	( )	( )
Root Bend	( )	( )

Instructor's initials: _____

Job 31-2

Fillet Weld Performance Test

Name _____ Date _____

Class _____ Instructor _____

> **Learning Objective**
> - In this job, you will perform a fillet weld performance test. Your instructor will select the type of material and process to be used.

1. Obtain two pieces of material. One piece should be 3/16″ × 4″ × 6″ (4.8mm × 100mm × 150mm), the other piece should be 3/16″ × 3″ × 6″ (4.8mm × 75mm × 150mm).

2. Obtain the correct electrode or filler metal for the base metal you will be welding.

3. Set up the welding equipment as appropriate for the material being welded.

4. Align the pieces to be welded as shown in the following figure.

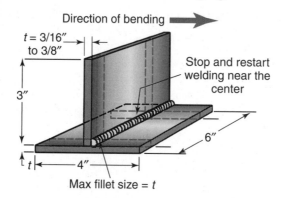

5. Tack weld the plates into this position.

6. Weld from one end toward the center. When you reach the center, stop welding. Restart your weld at the center and finish the fillet weld.

7. Clamp the base piece in a vise. Apply a load, using a long bar or other tool, so the vertical piece folds down over the weld. Continue to bend the vertical piece until the weld breaks or the two pieces touch one another.

    ■ **Caution:** Use approved eye protection while performing this test.

> **Inspection**
> Examine the exposed surface of the weld for lack of fusion, inclusions, pits, or other discontinuities. There should be very little or no evidence of these flaws.
>
> Instructor's initials: _____

Lesson 32

The Welding Shop

Name _____ Date _____

Class _____ Instructor _____

Learning Objective

● You will be able to identify and use various pieces of equipment in a welding shop.

Instructions

Carefully read Chapter 32 of the text and review the figures in the chapter. Then answer the following questions.

1. Which of the following is *not* a reason to use a ventilation system?
   A. Supply clean air and oxygen to workers.
   B. Keep gases and fumes to a minimum.
   C. Remove all toxic fumes from the welding shop.
   D. Heat the replacement air during cold weather.

1. _____

2. A(n) ____ shear is used for cutting thin metals.

2. _____

3. After cutting, metal is often formed. What machine is commonly used to make straight-edged bends in metal?

_____

4. List three types of preheaters.

_____

_____

_____

5. In a welding shop, jigs and fixtures are extensively used to ____.
   A. add additional heat to the parts being welded
   B. minimize the amount of filler metal used
   C. reduce spatter
   D. prevent the parts from warping or buckling during welding

5. _____

6. *True or False?* Air pressure is used to propel abrasives during a blast cleaning operation.

6. _____

7. Identify the following piece of equipment.

_____

8. Review the list of welding shop tools presented in section 32.4 of the text. Think about each tool and list six tools that you think are used most often by welders in industry.

_____

_____

_____

9. Performing a(n) _____ requires little or no equipment.
   A. visual inspection
   B. x-ray inspection
   C. hardness test
   D. bend test

9. _____

10. Which of the following is *not* a consideration in estimating the final cost of fabricating an article?
    A. Type and quantity of material to be used.
    B. Length, depth, and shape of the weld.
    C. Intended use of the finished weldment.
    D. Labor costs.

10. _____

Lesson 33

Getting and Holding a Job in the Welding Industry

Name _____ Date _____
Class _____ Instructor _____

Learning Objective

● You will be able to identify the information that you will need to prepare for a job interview and to complete a job application. You will also be able to list the attitudes that you should display during an interview and on the job.

Instructions

Carefully read Chapter 33 of the text and review the figures in the chapter. Then answer the following questions.

1. Assume that you are applying for a job as a gas metal arc welder in a large steel fabrication plant. To the best of your ability, complete the Application for Employment form presented in this lesson.

2. After completing the application, list all the items for which you did not have sufficient information.

   ■ **Note:** You should have these and other items of information noted on a piece of paper to take with you when you apply for a job.

   _____
   _____
   _____
   _____

3. How do you intend to enhance your personal appearance in preparation to job seeking and for a job interview?

   _____
   _____
   _____
   _____

Copyright by The Goodheart-Willcox Co., Inc. Modern Welding Lab Workbook

4. If asked what you hope to be doing three years from now, what would you answer? (Your answer to this question shows an employer that you have goals.)

_____
_____
_____

5. How would you respond if asked, "Why do you want to work for our company?"

_____
_____
_____
_____
_____
_____

6. Imagine that you are an employer. What would you think of someone you were interviewing who kept looking at the floor or could not look you square in the eye?

_____
_____
_____

7. If you did poorly in completing this lesson or the job application, ask yourself the following questions:
   A. "Am I able to follow written directions?" (Yes/No)
   B. "Am I able to express myself in writing?" (Yes/No)
   If your answer to either of these is "No", then you should work to gain these skills.

Lesson 33 Getting and Holding a Job in the Welding Industry

Name _____

APPLICATION FOR EMPLOYMENT

PERSONAL INFORMATION

Date _____

Name _____
          Last   First   Middle

Present Address _____
                    Street   City   State

Permanent Address _____
                      Street   City   State

Phone No. _____

If related to anyone in our employ,   Referred
state name and department   by

EMPLOYMENT DESIRED

Position _____ Date you can start _____ Salary desired _____

Are you employed now? _____ If so, may we inquire of your present employer? _____

Ever applied to this company before? _____ Where _____ When _____

EDUCATION

| | Name and Location of School | Years Completed | Subjects Studied |

Grammar School _____

High School _____

College _____

Other _____

Subject of special study or research work _____

(Continued)

Modern Welding Lab Workbook

What foreign languages do you speak fluently? _____ Read _____ Write _____

U.S. Military service _____ Rank _____ Present membership in National Guard or Reserves _____

Job-related activities and skills _____

_____

FORMER EMPLOYERS List below last four employers starting with last one first

Date Month and Year	Name and Address of Employer	Salary	Position	Reason for Leaving
From				
To				
From				
To				
From				
To				
From				
To				

REFERENCES Give below the names of three persons not related to you, whom you have known at least one year

	Name	Address	Business	Years Acquainted
1				
2				
3				

PHYSICAL RECORD

List any factors that would prevent you from being physically able to perform this job.

_____

In case of emergency notify _____ Name _____ Address _____ Phone No. _____

I authorize investigation of all statements contained in this application. I understand that misrepresentation or omission of facts called for is cause for dismissal.

Date _____ Signature _____

Lesson 34
Technical Data

Name _____ Date _____

Class _____ Instructor _____

Learning Objective
You will be able to understand and use the technical information necessary to complete a job in the welding shop. You will also be able to use and convert various SI Metric to US Customary units and vice versa.

Instructions
Carefully read Chapter 34 of the text and review the figures in the chapter. Then answer the following questions.

1. What three factors affect the volume of gas that reaches the torch valves?
   _____
   _____
   _____

2. What is the gauge reading of the pressure within an oxygen cylinder when the outside temperature is 40°F?   2. _____

3. What is the chemical formula for propane?   3. _____

4. What is the melting temperature of titanium in °C?   4. _____

5. What is the chemical symbol and melting temperature for chromium?   5. _____

6. What is the weight in pounds of four cubic feet of silver? (You will need the weight per cubic foot from Figure 34-4.) Show your calculations.   6. _____

7. What is the approximate temperature in °C at which clean carbon steel changes to an orange red color?

7. _____

8. What is the temperature of an oxy-propylene neutral flame?

8. _____

9. What decimal size is an "I" drill?

9. _____

10. What size tap drill should be used to drill and tap a hole for a 5/16 NC thread? (NC thread sizes have fewer threads than an NF thread sizes.)

10. _____

11. To drill a hole through which a 7/16" bolt would pass with only 1/64" clearance all around, what size factional, letter, or number drill should be used? Show your calculations.

11. _____

12. The SI Metric system base unit for mass is _____.
    A. kg
    B. K
    C. mol
    D. m

12. _____

13. The SI Metric system base unit for length is _____.
    A. K
    B. in
    C. A
    D. m

13. _____

14. How much would steel expand if it was heated from 70°F to 1000°F? (You will need the expansion per °F rise in temperature from Figure 34-5.) Show your calculations.

14. _____

Lesson 34 Technical Data

Name _____

15. Convert a preheating temperature of 800°C to °F. Show your calculations.

15. _____

16. If the brazing temperature of a metal is 820°F, what would the temperature be in °C? Show your calculations.

16. _____

17. If the pressure in a cylinder is 2000 psi, what is the pressure in Pascals (Pa)? Show your calculations.

17. _____

18. If water is flowing at the rate of 2,000 ft³/hr, how fast would it be flowing in liter/min? Show your calculations.

18. _____

19. Answer the following two questions for using a #7 cutting tip with a 100 foot hose with a 1/4" ID and had set the flowing gauge pressure at 100 psi.

   A. What would the pressure be at the cutting tip? _____

   B. How much pressure would have been lost over the 100 feet of hose? _____

20. What would the tensile strength of a metal sample be in kPa if   20. _____
    it is measured at 100,000 psi? Show your calculations.